PROCESS TECHNOLOGIES FOR PHOSPHATE FERTILIZERS

PROCESS TECHNOLOGIES FOR PHOSPHATE FERTILIZERS

UNITED NATIONS INDUSTRIAL DEVELOPMENT ORGANIZATION

University Press of the Pacific
Honolulu, Hawaii

Process Technologies for Phosphate Fertilizers

by
United Nations Industrial Development Organization

ISBN: 1-4102-0627-0

Copyright © 2003 by University Press of the Pacific

Reprinted from the 1978 edition

University Press of the Pacific
Honolulu, Hawaii
http://www.universitypressofthepacific.com

All rights reserved, including the right to reproduce this book, or portions thereof, in any form.

Preface

This volume and its companion in the Development and Transfer of Technology Series, *Process Technologies for Nitrogen Fertilizers,* are the first two documents to emanate from the Industrial and Technological Information Bank (INTIB) of the United Nations Industrial Development Organization (UNIDO), which is a component of the UNIDO programme on the development of technology. INTIB is a pilot operation that began in July 1977 for a period of 18 months. Its work is being concentrated on four industrial sectors: iron and steel, fertilizers, agro-industries, and agricultural machinery and implements. Each of these sectors has priority in other UNIDO endeavours also: sectoral studies, consultations, negotiations and technical assistance projects.

The concept of INTIB has its roots in the Lima Declaration and Plan of Action, adopted at the Second General Conference of UNIDO in 1975, and in various United Nations General Assembly resolutions, all envisaging such a service as an essential instrument for the transfer, development and adaptation of appropriate technologies. To increase the share of developing countries in world industrial output from 7 per cent (in 1975) to 25 per cent in 2000, an objective set by the Lima Conference, decision makers must have adequate information on new investments. Those advising the decision makers—national industrial information centres, technology development institutes and investment banks—also must have this information.

The novel character of INTIB, as compared with the services previously rendered by UNIDO, is that it is addressed to the selection of technology before the technology is acquired. INTIB not only draws upon the services available in the Industrial Information Section, where it is housed, but also relies on the expertise of specialists in the Industrial Operations Division of UNIDO and outside experts to process the information obtained from sources within and outside UNIDO relevant to technology selection. As a result of this effort, UNIDO is able to supply information in anticipation of, as well as in response to, demand. An example is the series of technology profiles and monographs being prepared, to which this volume belongs, concerning matters to consider when selecting a technology from a variety of alternatives.

The "customers" INTIB is designed to serve include ministries of industry, planning and industrial development institutes, multipurpose technological institutions, and agencies concerned with the transfer of technology. The intention is to serve all those who are responsible for selecting technology, whether in an advisory role or decision-making capacity, in each of the four priority industrial sectors selected for the pilot phase.

Further information about INTIB and its related activities can be had on request by writing to the Chief, Industrial Information Section, UNIDO, P.O. Box 707, A-1011 Vienna, Austria.

This description of phosphate fertilizer production technologies was prepared by K. R. Krishnaswami, acting as a consultant to UNIDO; the views expressed are those of the consultant and do not necessarily reflect the views of the secretariat of UNIDO.

EXPLANATORY NOTES

References to dollars ($) are to United States dollars.

The standard system of NPK fertilizer grades is used. For example, a 10-15-12 fertilizer is one containing, in available forms, 10% nitrogen, 15% phosphorus expressed as P_2O_5, and 12% potassium expressed as K_2O.

The units used are those of, or accepted for use with, the International System of Units (SI); in particular:

t	tonne (metric ton, 1 t = 1 000 kg)
t/d	tonne per day
bar	bar (1 bar = 10^5 Pa = 0.987 atm = 1.02 kgf/cm^2)

In addition, the following abbreviations have been used:

BPL	bone phosphate of lime ($Ca_3(PO_4)_2$)
rpm	revolution per minute
TVA	Tennessee Valley Authority

CONTENTS

	Page
INTRODUCTION	1
SULPHURIC ACID	3
Contact process–normal version	3
Contact process–high-pressure version	4
From pyrites	4
Müller-Kühne method	5
PHOSPHORIC ACID	7
Nissan hemihydrate-dihydrate process (conventional)	7
Fisons hemihydrate process	9
Classic Prayon dihydrate process	10
Central Glass/Prayon process (dihydrate-hemihydrate)	11
Gulf Design isothermal process (dihydrate)	12
Dorr Oliver high-yield high-strength (HYS) process	13
Hydrochloric acid extraction process	13
Electric furnace process	14
SUPERPHOSPHORIC ACID	17
Concentration of wet-process acid	17
Concentration of furnace acid	19
SINGLE SUPERPHOSPHATE (SSP)	20
TRIPLE SUPERPHOSPHATE (TSP)	22
MONOAMMONIUM PHOSPHATE (MAP)	24
Fisons process (powder)	24
PhoSAI process (powder)	25
Slurry process (granules)	25
DIAMMONIUM PHOSPHATE (DAP)	27
AMMONIUM SULPHATE PHOSPHATE	29
AMMONIUM PHOSPHATE NITRATE (WITH POTASH)	31
UREA AMMONIUM PHOSPHATE (UAP)	33
Conventional slurry process	33
Melt process	34
Dry-ingredients process	36
Melt oil-cooling process	37

	Page
NITROPHOSPHATES	39
Odda process	39
Sulphonitric process	41
Phosphonitric process	42
Carbonitric process	43
Annex — List of firms	45
Bibliography	49

Introduction

The purpose of this compendium of the most important processes used for making phosphate fertilizer materials is to provide a guide to the selection of process technologies for developing countries interested in initiating efforts in this sector. The processes for a given product are grouped together, and in every case the following information is given:

Description of the process operations

Flow chart

Advantages and disadvantages relative to the other processes in the same section

Owners, in the case of proprietary processes

Engineering licensees, if any

Cost of erecting a typical plant

Unless otherwise stated, the know-how fees for proprietary processes are about 10% of the f.o.b. or f.o.r. costs of plant and equipment. The estimates of the cost of erecting typical plants are based on 1977 prices.

More details about the processes can be found in the publications listed in the bibliography and from the firms named in the descriptions, whose addresses are given in the annex. The list of firms (whose names were supplied by the consultant) should not be regarded as exhaustive, and the omission of a given firm's name is no more a sign of disapproval than its inclusion would constitute a recommendation.

All the processes described here are based on the use of naturally occurring phosphate rock as the prime raw material. The different steps in the processing of phosphate rock adopted in fertilizer manufacture are shown in a generalized way in the following diagram:

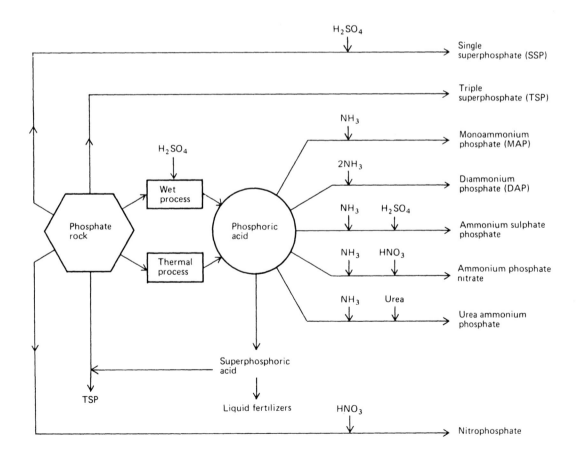

The predominant phosphate minerals in the rock are fluorapatite, $3Ca_3(PO_4)_2 \cdot CaF_2$, and hydroxyapatite, $3Ca_3(PO_4)_2 \cdot Ca(OH)_2$. They supply the bulk of the phosphorus component in fertilizers manufactured all over the world. However, the powdered rock itself is useless as a fertilizer because the phosphorus in it is not in the water-soluble form that would make it readily available to plants. Reacting the rock with sulphuric acid (H_2SO_4), nitric acid (HNO_3) or hydrochloric acid (HCl) leads to phosphoric and superphosphoric acids (H_3PO_4), important intermediates in the production of phosphate fertilizers. Because of its special advantages, the reaction with sulphuric acid is the one most often used.

The combination of nitrogen (N) and phosphorus (P) in NP fertilizers is achieved by specially controlled processes involving the addition of such nitrogen compounds as ammonia (NH_3), urea (NH_2CONH_2) and nitric acid to phosphoric acid. Fertilizers also containing the third primary nutrient, potassium (K), are made by the physical addition of compounds like potassium chloride (KCl) and potassium sulphate (K_2SO_4) to NP fertilizers and granulating the mixture to make it homogeneous. Some grades of such NPK mixtures are also made by bulk-blending several individual fertilizers containing one or more of the nutrients.

Because of the importance of sulphuric acid in the manufacture of phosphoric acid, and of ammonia in providing nitrogen in fertilizer materials, the manufacture of sulphuric acid is described in this volume and that of ammonia in the companion volume, *Process Technologies for Nitrogen Fertilizers*.

Sulphuric acid

The three methods described here for the manufacture of sulphuric acid (H_2SO_4) start from different raw materials:

(a) Elemental sulphur, including sulphur recovered by the Frasch process, refined native sulphur, or recovered sulphur from other sources (e.g. petroleum refineries). The most important processes used in the contact process;

(b) Iron sulphide ores (pyrites), by roasting;

(c) Calcium sulphates, including natural gypsum, by the Müller-Kühne method.

Contact process—normal version

The elemental sulphur used as raw material in the contact process is 99.5% pure and free from arsenic, chloride and fluoride; it contains less than 0.2% and 0.25% carbon. Arriving at the plant in molten or lump form, it is burnt in an excess of dry air. Sulphur dioxide is formed:

$$S + O_2 \rightarrow SO_2 \qquad (1)$$

The gases containing SO_2 are cooled in a waste-heat boiler to 400°-600°C and then passed over a vanadium pentoxide catalyst charged in a converter, where the SO_2 is oxidized:

$$SO_2 + [O] \rightarrow SO_3 \qquad (2)$$

The gases are cooled and sent to an absorption tower. Sulphuric acid of 98% strength is recirculated in the tower and SO_3 is absorbed to produce strong acid:

$$SO_3 + H_2O \rightarrow H_2SO_4 \qquad (3)$$

Water is added to the strong acid to restore the original strength, and product acid is withdrawn. Unabsorbed gases are vented to the atmosphere.

Various attempts have been made to increase the conversion of SO_2 to SO_3. In one of these, the double catalyst/double absorption (DC/DA) system, the process flow is modified so that after the second or third pass in the converter the gases are withdrawn and passed through the absorber to remove the SO_3.

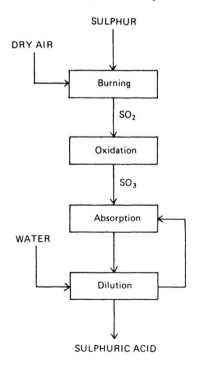

Sulphuric acid from sulphur by the contact process

The remaining gases are reheated and sent through the last pass over the catalyst; further absorption follows. An increased conversion, up to 99.5%-99.7%, is obtained. Vanadium pentoxide is a good, standardized catalyst for SO_2 conversion that can improve the conversion up to 99.9% in DC/DA plants.

Advantages

Enhanced conversion of SO_2 to SO_3 at 99.9%
Possible higher inlet SO_2 concentration, thus increasing plant capacity to some extent
Anti-pollution standards for stack gases are met because of the low concentration of unabsorbed sulphur oxides in the stack

Disadvantages

High initial plant cost

Process owners

Farbenfabriken Bayer AG
Lurgi Chemie und Hüttentechnik GmbH
PEC-Engineering
Produits Chimiques Pechiney-Saint-Gobain
Ugine Kuhlmann SA

Engineering licensees

Chemical Construction Corporation (Chemico)
Davy Powergas GmbH
Davy Powergas Ltd
Foster Wheeler Corporation
Friedrich Uhde GmbH
Heurtey Industries
Humphreys and Glasgow Ltd
Krebs et Cie SA
Lurgi Chemie und Hüttentechnik GmbH
Monsanto Enviro-Chem Systems, Inc.
Nissan Chemical Industries Ltd
Sim-Chem Division, Simon-Carves Ltd

Cost of typical plants

A 500 t/d plant will cost $8 million, including sulphur handling and storage facilities. A 1 000 t/d plant can cost $12 million to $14 million and a 1 500 t/d plant $15 million to $18 million. The cost can be higher by 15% for plants using the DC/DA system.

Contact process—high-pressure version

Air is compressed to 7 bar, dried and recompressed to 22 bar before admission into the furnace. Exit gases from the converter are passed through a waste-heat boiler, a converter, an economizer and an absorber.

Advantages

High conversion, up to 99.8%
Less sulphur oxides to stack

Disadvantages

Extra energy required for compression, although a part of it can be recovered in an expansion turbine

Process owners

Ugine Kuhlmann SA

Engineering licensees

Canadian Industries Ltd

Cost of typical plant

A 650 t/d plant costs about $14 million to $16 million.

From pyrites

Large tonnages of sulphuric acid are produced from pyrite (FeS_2) and pyrrhotite (FeS). These ores contain 45%-48% and 30%-32% sulphur, respectively.

The ore is crushed and roasted in a vertical multiple-hearth furnace fitted with rakes. Alternatively, it can be roasted in a fluidized-bed calciner. Dust and other impurities like chlorine, fluorine and arsenic are removed from off-gases by precipitators and scrubbers. SO_2 is converted to SO_3 and absorbed to H_2SO_4 by the contact process (see above).

By-product iron oxide from the roasting is sold for iron and steel manufacture, reducing the cost of manufacturing sulphuric acid.

Advantages

Use of elemental sulphur avoided

Disadvantages

High initial plant cost

Process owners

BASF Aktiengesellschaft
Montedison SpA
Outokumpu Oy

Engineering licensees

AB Celleco
Chemical Construction Corporation (Chemico)
Haines and Associates, Inc.
Jacobs Engineering Company
The Lummus Company

Cost of typical plant

A 1 000 t/d plant can cost about $20 million.

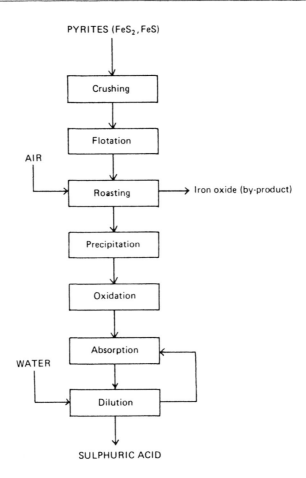

Sulphuric acid from pyrites

Müller-Kühne method

Sulphuric acid is manufactured from calcium sulphate ($CaSO_4$) by the Müller-Kühne method, which involves the reactions

$$CaSO_4 + 2C \rightarrow CaS + 2CO_2 \quad (1)$$

$$3CaSO_4 + CaS \rightarrow 4CaO + 4SO_2 \quad (2)$$

and can be represented overall by the equation

$$4CaSO_4 + 2C \rightarrow 2CO_2 + 4CaO + 4SO_2$$

The raw material can be natural gypsum ($CaSO_4 \cdot 2H_2O$), natural anhydrite, ($CaSO_4$) or gypsum obtained as a by-product in phosphoric acid manufacture (see below). Reaction 1, producing the intermediate CaS, takes place at 700°C and reaction 2 at 900°C. Calcining is carried out in a rotary kiln. The CaO reacts with the additives, clay ($Al_2O_3 \cdot SiO_2$) and Fe_2O_3; as a result, clinker is produced. After addition of gypsum as a retarder, the clinker is ground to cement. The SO_2 liberated during the reaction is purified and converted to sulphuric acid by the contact process (see above).

Advantages

Possible to use a variety of raw materials: anhydrite, natural or by-product gypsum
Process mostly free from elemental sulphur requirement
Low cost of manufacturing

Disadvantages

High initial cost
Quality of raw material critical if by-product gypsum is used

Process owners

Chemie Linz AG
Rhone-Poulenc Group

Engineering licensees

Büttner Scheide Haas GmbH
Fried. Krupp GmbH
Friedrich Uhde GmbH

Krebs et Cie SA
Sim-Chem Division, Simon-Carves Ltd
Société de Prayon
Swindell Dressler Company
VEB Chemieanlagenbau Magdeburg

Cost of typical plant

The investment cost will be about double that of a plant burning sulphur. For a 1 000 t/d plant, the cost will be about $28 million.

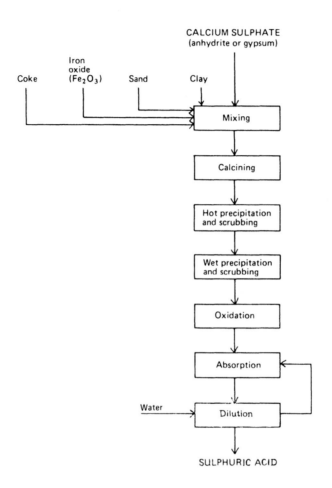

Sulphuric acid from calcium sulphate by the Müller-Kühne method

Phosphoric acid

Phosphoric acid (H_3PO_4), the key intermediate in the manufacture of all phosphate fertilizers except single superphosphate and the nitrophosphates, is manufactured from phosphate rock. Process routes for making the acid can be classified as wet or thermal; the thermal processes give an acid of higher purity.

In a wet process the phosphate rock is digested with an acid. The reaction taking place with sulphuric acid produces gypsum and hydrogen fluoride as by-products:

$$Ca_{10}(PO_4)_6F_2 + 10H_2SO_4 + 20H_2O \rightarrow 10CaSO_4 \cdot 2H_2O + 6H_3PO_4 + 2HF$$

The reaction itself is fast, but the gypsum crystals form rather slowly and P_2O_5 tends to be included in the lattice.

The various commercial processes developed have these basic objectives:

(a) Transferring the maximum amount of P_2O_5 from the rock to the product acid (or in other words, ensuring that as little P_2O_5 as possible is lost in the gypsum cake);

(b) Obtaining as high a percentage of P_2O_5 in the product acid (45%-50% P_2O_5) as possible;

(c) Having by-product gypsum in a suitable form for filtration and of acceptable quality for further use.

Seven wet processes are described, one using hydrochloric acid.

The most important thermal method available for converting phosphorus in the rock to phosphoric acid is the electric furnace process. In this process, the furnace is charged with phosphate rock, coke and silica particles. When the charge is heated, the following reaction takes place:

$$2Ca_3(PO_4)_2 + 6SiO_2 + 10C \rightarrow 6CaSiO_3 + P_4 + 10CO$$

The rest of the process is similar to the contact process for making sulphuric acid: The elemental phosphorus is burned and its oxides absorbed in water to make the acid.

The last process described in this section is the electric furnace thermal process.

Nissan hemihydrate-dihydrate process (conventional)

Phosphate rock (usually 75% BPL, which is typical for rock from Morocco) ground to such a size that 90% passes through 100 mesh and sulphuric acid diluted to 60% strength are fed at predetermined rates to the first of two digesters provided with agitators. The ratio between the weights of rock and sulphuric acid (as 100% H_2SO_4) is 1.1-1.2 depending on the rock quality. Weak phosphoric acid from the filtration section (back end of the plant) is also recycled to the first digester, and slurry from the first digester flows to the second one. A temperature of 90°-100°C is maintained in the digesters. These conditions are favourable for the growth of calcium sulphate hemihydrate ($CaSO_4 \cdot \frac{1}{2}H_2O$) crystals.

The hemihydrate slurry from the second digester flows to three crystallizers in series where hydrated crystals are already present. Cooling is done with air to maintain a temperature of 60°-65°C. Under these conditions the hemihydrate crystals are dissolved so that the phosphate inclusions are transferred to the acid. Sufficient time is allowed for the seed crystals to grow, and the slurry containing the dihydrate (gypsum, $CaSO_4 \cdot 2H_2O$) is sent to filtration or, in part, is recycled to the first crystallizer to maintain optimum conditions. The filtration is done in a tilting-pan filter, and the product acid, which is the filtrate, has a concentration of 30%-32% P_2O_5. This filtrate is sent to evaporators for increasing the concentration to 45%-54%. The filtered cake is subjected to two countercurrent washes in the filter before it is sent to the battery limits for disposal or

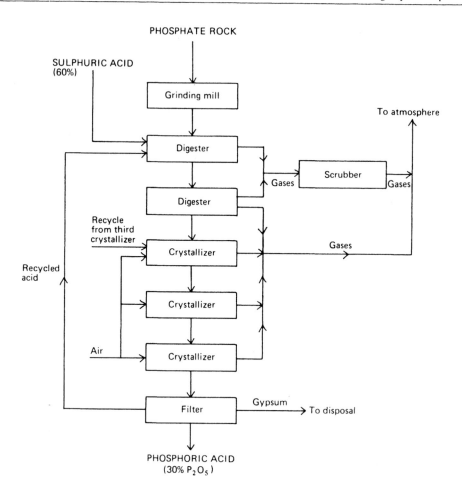

Phosphoric acid by the Nissan hemihydrate-dihydrate process (conventional)

for further use. Effluent gases from the digesters are washed in a scrubber to remove fluorine, and the scrubbed gases are vented to the atmosphere along with the gases from the crystallizers.

Advantages

    Process well proved, and in operation in more than 20 plants throughout the world
    Gypsum slurry with good filtration qualities and minimal scaling problems
    Good quality gypsum cake with P_2O_5 content of 0.3%
    P_2O_5 recovery efficiency up to 98%

Disadvantages

    Rock grinding (consuming about 15 kWh per tonne of rock) and further concentration of the product acid of 30%-32% in most cases required, which means additional capital and operating costs

Process owners

    Fisons Ltd, Fertilizer Division
    Mitsubishi Chemical Industries Ltd
    Nissan Chemical Industries Ltd

Engineering licensees

    Friedrich Uhde GmbH
    Matthew Hall Engineering Ltd
    PEC-Engineering
    Sim-Chem Division, Simon-Carves Ltd

Cost of typical plant

The investment cost, including rock grinding, vacuum-concentration and necessary off-site facilities, but excluding sulphuric acid plant, for a 150 t/d P_2O_5 plant (about 280 t/d of 54% P_2O_5 acid) will be about $8 million. The 300 and 600 t/d plants on a similar basis can cost $13 million and $17 million. A 2 x 1 000 t/d plant can cost $43 million to $44 million.

Fisons hemihydrate process

Unground phosphate rock (usually 73%-75% BPL), 100% passing through 10 mesh, is fed to the first of the two reactors. There it is mixed with slurry recirculating from the second reactor and passing through a flash cooler to control the temperature. The slurry overflows from the first to the second reactor. Sulphuric acid of 93% strength is fed to the second reactor. Both reactors are provided with bladed agitators and are suitably baffled inside to ensure effective mixing. The temperature in the reactors is kept at 98°-102°C, and the free SO_4 in the second reactor is maintained at 1.8%-2.2% by adjusting the sulphuric-acid and return-acid flow from the filter to this reactor. Slurry overflows from the second reactor to the filter feed-tank. It is then pumped to a horizontal tilting-pan vacuum filter to remove the hemihydrate ($CaSO_4 \cdot \tfrac{1}{2}H_2O$) crystals. Product acid of 50% P_2O_5 strength is drawn off from the first-stage filtration; part of this acid is recycled to the second reactor along with filtrate from the subsequent stages; the remainder of the acid is stored for further use.

The gases evolving from the reactors and the filter feed-tank are washed in a scrubber to remove the fluorides before discharge to the atmosphere, along with the scrubbed off-gases from the flash cooler.

Advantages

High product-acid concentration, between 44% and 52% P_2O_5; no further concentration needed and can be used for further process directly
Unground coarse rock acceptable
Lower initial cost of plant

Disadvantages

Low P_2O_5 recovery efficiency, at 93%-94%, a factor that increases the operating cost

Process owners

Fisons Ltd, Fertilizer Division

Engineering licensees

Davy Powergas Ltd
Foster Wheeler Corporation
Humphreys and Glasgow Ltd
Lurgi Chemie und Hüttentechnik GmbH

Cost of typical plant

The investment cost, excluding sulphuric acid plant, for a 150 t/d P_2O_5 capacity plant (about 280 t/d of 54% P_2O_5 acid) will be $7 million. The 300 and 600 t/d plants on a similar basis can cost around $12 million and $15 million. A 2 x 1 000 t/d plant can cost around $38 million to $40 million.

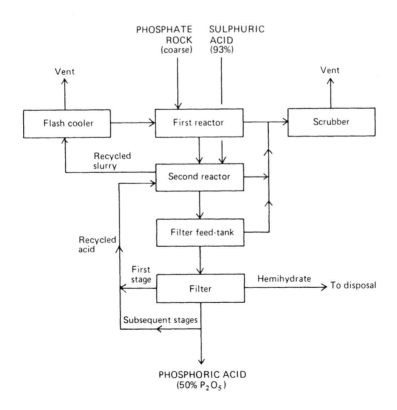

Phosphoric acid by the Fisons hemihydrate process

Classic Prayon dihydrate process

In the classic Prayon process, rock-grinding mills, reactor, sulphuric acid cooler, filter, flash cooler, recycle system for slurry and fume scrubber are the major pieces of equipment. The reactor has many compartments, separated by baffles and provided with agitators. Phosphate rock (usually 75% BPL), ground to such size that 60% passes through 200 mesh, is admitted to the first compartment. Dilute sulphuric acid of 50%-60% strength is fed to the first three compartments. The slurry flows from compartment to compartment either over or under the baffles. The temperature in the reactor is maintained at 70°-80°C by flash cooling the slurry from the penultimate compartment and recycling part of it to the first compartment. The remaining slurry goes to the hold tank and then to filtration, which is the conventional Prayon tilting-pan type. The product acid of 30%-32% is collected from the first filter section and sent to the evaporator for concentration to 45%-54%. The gypsum slurry is sent for disposal.

The classic Prayon process has been improved recently by providing for better agitation, a better vacuum manifold for the Prayon filter and the possibility of using coarser rock and concentrated acid directly (thereby eliminating the cooler). This process, called the convertible Prayon process, is also a dihyrate process. In it, the gypsum quality can be improved by adding a few pieces of equipment at small extra cost; this step, it is claimed, will also increase plant capacity by 25%.

Advantages

    Simple and well proven process
    Low initial cost

Disadvantages

    Low product acid concentration, 28%-32%
    Low P_2O_5 recovery efficiency, 95%-96%
    Gypsum quality not good for wallboard or cement manufacture

Process owners

    Jacobs Engineering Company
    Produits Chimiques Pechiney-Saint-Gobain
    Société de Prayon SA

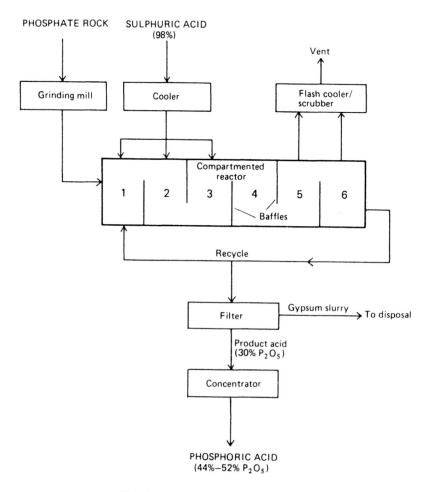

Phosphoric acid by the Prayon dihydrate process

Engineering licensees

    Coppée-Rust SA
    Davy Powergas, Inc.
    The Fertilizers and Chemicals, Travancore Ltd
    Sim-Chem Division, Simon-Carves Ltd

Cost of typical plant

The investment cost, including rock grinding, concentration and necessary off-site facilities but excluding sulphuric acid plant, for a 150 t/d P_2O_5 capacity (about 280 t/d of 54% P_2O_5 acid) will be about $8 million. The 300 and 600 t/d plants on a similar basis can cost $13 million and $17 million. A 2 × 1 000 t/d plant can cost about $42 million.

Central Glass/Prayon process
(dihydrate-hemihydrate)

The Central Glass/Prayon process is unique as it follows the route dihydrate recrystallized to hemihydrate. A multicompartment acidulator reactor, rock-grinding mills, a centrifuge, hemihydrate crystallizers, a Prayon filter, a flash cooler and a concentrator are the major equipment items. Phosphate rock (75% BPL), moderately ground (60% passing through 100 mesh), is acidulated with 98% sulphuric acid in the reactor. The temperature in the reactor is controlled at 70°-80°C by flash evaporation of the slurry. The slurry from the reactor is centrifuged, and the product acid of 30%-35% is drawn off and sent for evaporation to increase its strength to 48%-52%. The cake from the centrifuge, which is the dihydrate $CaSO_4 \cdot 2H_2O$, is sent to the recrystallizer, where additional sulphuric acid is added with heating steam. The acid excess of 10%-15% and the temperature of 85°C enable the hemihydrate $CaSO_4 \cdot \frac{1}{2}H_2O$ to crystallize. The slurry is sent to the Prayon filter. The wash liquor from this filter as weak acid is returned to the first-stage reaction to improve P_2O_5 recovery. The hemicake from the filter is sent for disposal or further use.

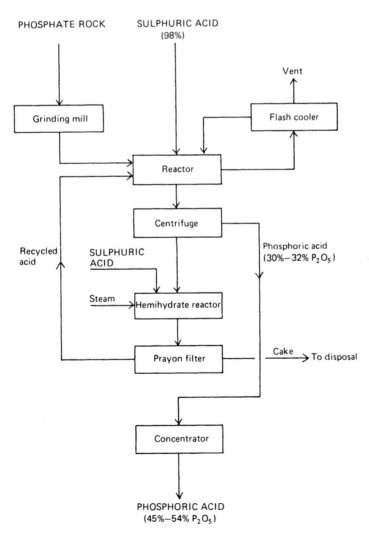

Phosphoric acid by the Central Glass/Prayon process

Advantages

    High P_2O_5 recovery efficiency of 98%-99%
    Gypsum of good quality for further use

Disadvantages

    Further concentration of product acid of 33%-35% needed
    Increased capital cost
    Some rock grinding necessary
    Limited operating experience with different kinds of rock

Process owners

    Société de Prayon SA

Engineering licensees

    Coppée-Rust SA
    Davy Powergas, Inc.
    The Fertilizers and Chemicals, Travancore Ltd
    Sim-Chem Division, Simon-Carves Ltd

Cost of typical plant

The investment cost, including rock grinding, concentration and necessary off-site facilities but excluding sulphuric acid plant, for a 150 t/d P_2O_5 capacity plant (about 280 t/d of 54% P_2O_5 acid) will be $8 million to $9 million. The 300 and 600 t/d plants on a similar basis can cost $14 million and $18 million. A 2 x 1 000 t/d plant will cost about $44 million.

Gulf Design isothermal process (dihydrate)

The Gulf Design isothermal process, developed in the last 12-15 years, uses as major equipment a pre-mix tank, an isothermal single reactor of the draft-tube type, a vacuum system, a filter feed-tank and a filter.

Unground phosphate rock (Florida, 72%-75% BPL) is mixed with return acid from the filter in the pre-mix tank and pumped as a slurry into the isothermal reactor at the bottom. The reactor consists of a draft tube at its centre and a high-capacity slurry-circulation arrangement, which, rotating at about 40 rpm, sends the slurry up the draft tube. Fresh sulphuric acid of 93% strength is sprayed above the slurry level in the reactor. The entire reactor is kept under vacuum by a steam-ejector system, and thus the circulating slurry is constantly exposed to the vacuum atmosphere. The heat of reaction generated in the reactor and the heat of dilution of the acid are removed through the vacuum system continuously. With this reactor-circu-

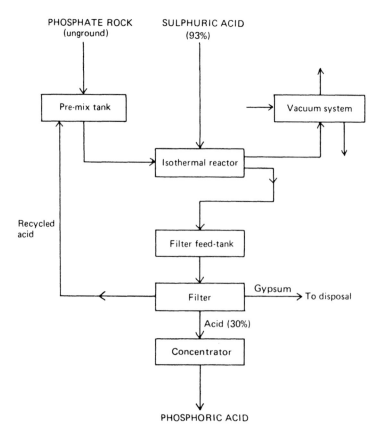

Phosphoric acid by the Gulf Design isothermal process

lation system, the maximum temperature difference in the entire reactor vessel is kept, it is claimed, as low as 0.3°C. The slurry from the reactor, tapped off near the outlet level of the draft tube, flows to the filter feed-tank and the filter. Product acid of about 30% concentration is taken to the concentration section and to storage, and the weak acid is returned to the pre-mix tank, while gypsum is sent for disposal.

Advantages

    Low capital cost
    Low power requirement
    Little atmospheric pollution

Disadvantages

    Low product-acid concentration, about 30% P_2O_5; further concentration needed
    Experience with different rocks limited
    Gypsum quality not good enough for wallboard or cement manufacture

Process owners

    Whiting Corporation, Swenson Division

Engineering licensees

    Gulf Design Company, Division of the Badger Company, Inc.

Cost of typical plant

The investment cost, including rock grinding, concentration and necessary off-site facilities but excluding a sulphuric acid plant, for a 300 t/d plant will be around $11 million, and a 600 t/d plant may cost around $15 million. A 2 x 1 000 t/d plant can cost $35 million to $38 million.

Dorr Oliver high-yield high-strength (HYS) process

The Dorr Oliver process consists of rock digestion in multitank vessels at high temperature with 93%-98% sulphuric acid and under elevated P_2O_5 concentration; under these conditions hemihydrate ($CaSO \cdot \frac{1}{2}H_2O$) crystals form. The temperature is maintained around 80°-85°C by a flash cooler and the recycling of the filtrate from the second-stage (dihydrate-stage) filtration. The concentrated acid is separated in the first-stage filter without washing the cake. The hemihydrate crystals are transferred to recrystallization reactors where sulphuric acid is added to solubilize the co-precipitated P_2O_5 in the cake. This reactor is either a multitank or a single-tank system. The heat of hydration is removed by flash cooling. The slurry is then pumped to the second-stage filter. The filtrate is recycled to the first-stage reactor, while the dihydrate cake is dumped out.

A major improvement in the equipment used in the process is the double dump filter, which filters the hemihydrate cake in one part of the cycle and the dihydrate in the other. After dihydrate removal, the empty pan is backwashed with high-pressure water to clean the cloth and remove the scale.

Dorr has introduced in this process addition of rock phosphate to the product acid to precipitate out the free acid. The product acid is permitted to settle and the clean acid sent for further use. The settled sludge is returned to the hemihydrate section to recover P_2O_5.

Advantages

    High strength acid (45% P_2O_5)
    High P_2O_5 recovery efficiency, 98.0%-98.5%
    Good-quality gypsum for wallboard and cement manufacture

Disadvantages

    High initial plant cost
    Limited operating experience with rocks other than Kola apatite
    Problems in filtration section for the hemihydrate cake reported

Process owners

    Dorr Oliver, Inc.
    Fisons Ltd, Fertilizer Division
    Kemira Oy
    Nissan Chemical Industries Ltd
    Singmaster and Breyer

Engineering licensees

    Jacobs Engineering

Cost of typical plant

The investment cost, including rock grinding, vacuum concentration and necessary off-site facilities but excluding sulphuric acid plant, for a 150 t/d P_2O_5 plant (about 280 t/d of 54% P_2O_5 acid) will be about $11 million. The 300 and 600 t/d plants on a similar basis can cost $14 million and $18 million. A 2 x 1 000 t/d plant can cost $46 million.

Hydrochloric acid extraction process

When phosphate rock is reacted with hydrochloric acid, a mixture of phosphoric acid and calcium chloride is obtained:

$$Ca_{10}(PO_4)_6F_2 + 20HCl \rightarrow 6H_3PO_4 + 10CaCl_2 + 2HF$$

Since both the phosphoric acid and the calcium chloride are in a dissolved state, physical means like filtration, centrifuging and crystallization cannot be used to separate them. Hence solvent extraction is

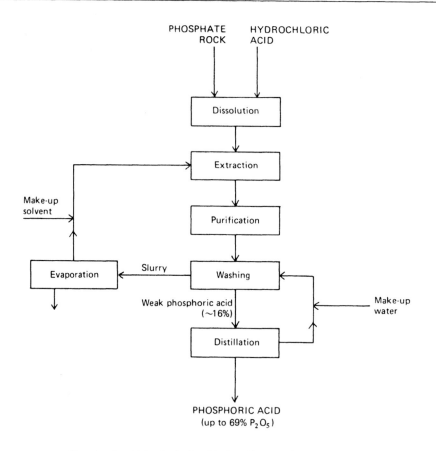

Phosphoric acid by the hydrochloric acid extraction process

adopted. Solvents that are commonly applied are butyl, amyl or isoamyl alcohol.

Phosphate rock and concentrated hydrochloric acid are mixed and the resulting solution sent to the extraction vessel where the solvent is also fed. Most of the phosphoric acid, excess hydrochloric acid and some calcium chloride are transferred to the solvent. Some of the calcium chloride present is removed in the following purifier, from which the liquor goes to the washing stage; here, water is used to extract the phosphoric acid from the liquor. The slurry goes to the evaporator, where the solvent is distilled off. The distillate solvent leaving this stage is virtually free from acid. The aqueous solution emerging from the washing stage contains H_3PO_4 (16% P_2O_5), HCl and a little of the solvent. This solution is concentrated to 69% P_2O_5 by distillation to yield the product. Hydrochloric acid and the bulk of the water boil off and return to the washing stage. The acid-free solvent from the evaporation is condensed and recycled to the extractor.

Advantages

High purity acid, almost comparable to thermal process acid, obtained
Hydrochloric acid used instead of sulphuric acid
High P_2O_5 recovery efficiency of 98%

Disadvantages

Low concentration of acid obtained—about 16%; further concentration needed
Adjasent chlorine-caustic plant required for hydrochloric acid
Operating experience not extensive

Process owners

IMI—Institute for Research and Development

Engineering licensees

Davy Powergas, Inc.

Cost of typical plant

The investment cost for a 150 t/d P_2O_5 plant, including concentration and off-site facilities but excluding hydrochloric acid plant, can be $8.0 million. The 300 and 600 t/d P_2O_5 plants on a similar basis can cost $13 million and $17 million. A 2 x 1 000 t/d plant will cost $42 million.

Electric furnace process

An electric furnace is charged with ground phosphate rock, coke and silica. The charge is made such that it has a SiO_2/CaO ratio between 0.8 and 1.2, and a P_2O_5/C ratio between 2.3 and 2.6. Only particles of correct size are charged to the furnace so as to have a porous bed. If necessary, phosphate rock is ground to 75% passing through a 0.1-mm mesh, and pellets are formed with this material using a binder.

The furnace can be of two types, either rectangular with three carbon electrodes in line, or triangular, again with three carbon electrodes. The gaseous products evolved during the melting of the mass are passed through precipitators and the water-cooled condenser, where phosphorus vapour is liquefied. The uncondensed gases, mostly carbon monoxide, are used as fuel in the rock-drying section.

The liquid phosphorus thus obtained is burnt in the combustion chamber with air, and the gases are cooled, hydrated to phosphoric acid and collected. The combustion of phosphorus and subsequent absorption can be carried out in one vessel, wherein the flame in the combustion chamber is surrounded by phosphoric acid flowing down the walls. Some of the resulting P_2O_5 collects into the acid, which is thus concentrated during the flow down the wall to the acid sump; fresh water is added to the sump to adjust the concentration. In other processes, combustion gases flow to a gas cooler, where the gases are cooled to about 180°C by a water spray. The gases then flow to the hydrator, where water and weak phosphoric acid are sprayed to form the product acid, which has a concentration up to 69%.

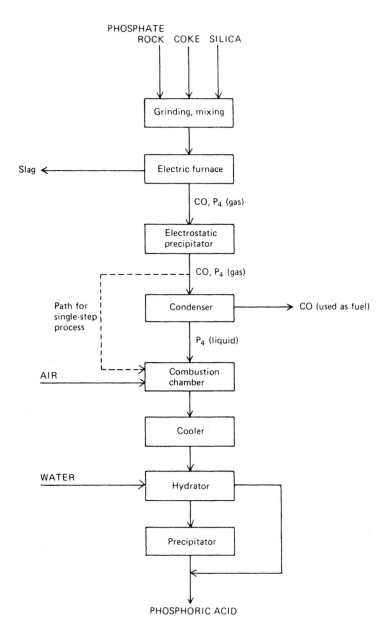

Phosphoric acid by the electric furnace process

Advantages

    High-purity acid with low dissolved solids and metal content
    High concentration of P_2O_5
    Acid suitable for manufacture of fine chemicals
    Rock quality not critical

Disadvantages

    Plant economically viable only in large capacities of 500 t/d and greater
    Electrical power requirement is high
    High capital cost

Process owners

    Albright and Wilson Ltd
    Kaltenbach et Cie SA
    Montedison SpA
    Pierrefitte-Auby SA
    Tennessee Valley Authority

Engineering licensees

    COCEI
    Davy Powergas Ltd
    Foster Wheeler Corporation
    Friedrich Uhde GmbH
    Gulf Design Company, Division of the Badger Company Inc.
    H. K. Ferguson Company, Inc.
    Kaltenbach et Cie SA
    The Lummus Company
    Lurgi Chemie und Hüttentechnik GmbH
    Monsanto Enviro-Chem Systems, Inc.
    Occidental Chemical Company
    Société Produits Chemiques et Synthesis SA

Cost of typical plant

A 200 t/d P_2O_5 plant can cost $18 million to $20 million, including rock handling, storage and phosphorus burning; a 300 t/d plant can cost $25 million to $27 million and a 2 x 300 t/d plant, $38 million to $42 million.

Superphosphoric acid

Superphosphoric acid contains 72%-83% P_2O_5, as compared with 54% P_2O_5 in wet-process phosphoric acid. It thus offers the advantage of lower freight costs, an important factor if shipments are to go overseas. Superphosphoric acid has good sequestering properties to keep the iron and alumina-containing sludge and also certain micronutrients like magnesium, copper and zinc in solution.

Superphosphoric acid from wet-process acid is used to make high-analysis liquid fertilizers of 8-27-0 grade. When it is reacted with phosphate rock, triple superphosphate containing 54% P_2O_5 can be produced. Triple superphosphate can be manufactured as granules, and almost all the P_2O_5 will be water soluble.

Superphosphoric acid is produced by dehydrating phosphoric acid made by either the wet-process or the electric furnace process. From wet-process acid, a final concentration of 68%-72% can be obtained, and with furnace acid, it can be 75%-83%. Two methods of the first kind will be described here, submerged combustion and vacuum concentration. The method described for concentrating furnace acid was developed by the Tennessee Valley Authority (TVA) and can be used without payment of know-how fees.

Concentration of wet-process acid

In the submerged combustion method, hot gases obtained by burning fuel in a furnace are made to flow into an evaporator fed with wet-process phosphoric acid. The concentrated acid is drawn off from an evaporator and cooled before it is sent to the sump and then to storage. There can be variations in the burner design. Either the hot gases from the furnace are passed through the wet acid or the flue gases from the burner may immediately be contacted with the acid. The product acid is transferred to storage. The off-gases from the evaporator pass through separators to remove entrained liquid. A

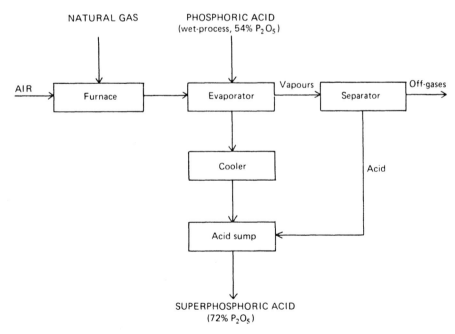

Superphosphoric acid by submerged combustion

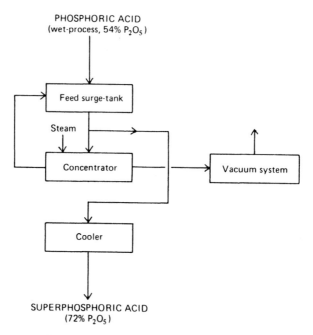

Superphosphoric acid by vacuum concentration

minimum throughput should be maintained in the evaporator so as to prevent solids from forming in the combustion area and at the same time to reduce undue entrainment losses.

Two processes involving vacuum concentration are in commercial use, one using a falling-film, the other a forced-circulation evaporator.

In the former, clarified acid (54% P_2O_5) is pumped to the top of the tube bundle of the falling-film evaporator. A set of steam ejectors maintains a vacuum in the tube side. The liquid is distributed across the tubes and is vaporized by the steam condensing on the shell side as the liquid film moves down. The concentrated acid flows into a separator and then to a recycle tank. The product is drawn off from this tank. The rest of the liquid is recirculated after mixing with fresh make-up liquid.

The forced-circulation evaporator system consists of a tube bundle heated on the shell side, a condenser vessel and an ejector system to maintain a vacuum. The feed acid after clarification is pumped to the tube side of the heat exchanger along with the recycle from the condenser separator vessel. Sufficient hydrostatic head is maintained on the submerged tubes to prevent boiling inside the tubes. As the level of the liquid rises through the tube bundle and enters the vapour head, the liquid begins to boil. The vapours rise to the surface, where they are disengaged and flow to the condenser. The separated particles of acid fall to the bottom of the condenser and then are taken by the recirculating pump. The product acid is tapped from this stream. The recycle is then made up with fresh acid and sent to the tube bundle.

Advantages

Product acid of high concentration, up to 72% P_2O_5

Lower losses owing to entrainment in vacuum evaporation system, as compared with submerged combustion system

Ease of fume scrubbing in vacuum evaporation system

Disadvantages

Possible problems arising from formation of solids, corrosion and scaling that will limit plant capacity in submerged combustion system

Plant capacity heavily dependent on vacuum attained from the vacuum evaporation system

Process owners

Albright and Wilson Company
Azote et Produits Chimiques (APC) SA
Produits Chimiques Pechiney-Saint-Gobain
Struthers Scientific and International Corporation
Swenson Evaporator Company
Tennessee Valley Authority

Engineering licensees

A.P.V. Kestner Ltd
Collier Carbon and Chemical Corporation

Davy Powergas Ltd
Gulf Design Company
Hitachi Zosen
The M. W. Kellogg Company
Occidental Chemical Company
Société de Prayon SA
Struthers Scientific and International Corporation
Submerged Combustion (Engineering) Ltd
Wellman Incandescent Ltd
Woodall-Duckham Ltd

Cost of typical plant

A 100 t/d plant can cost about $1.4 million and a 200 t/d plant can cost $2.0 million.

Concentration of furnace acid

In the furnace process the product can be concentrated to superphosphoric acid in the hydration step itself, when the combustion chamber off-gases from phosphorus burning are treated with water. In practice it is done by decreasing the water supply to the hydrator; however, since the hydrator temperature would go up by this step, there is a limit beyond which the water supply cannot be decreased. This limit largely depends on the cooling capacity provided in the hydrator acid-recirculation system.

Information published by TVA describes this system, which mainly consists of combustion chambers, hydrator, product-acid cooler and a venturi scrubber followed by a separate tower. The critical equipment is of 316 ELC stainless steel. The combustion chamber is an unlined vessel, cooled externally by flowing water. Air for combustion is supplied by centrifugal blowers. During the combustion, a layer of solid or semisolid polyphosphoric acid forms on the interior of the combustion chamber and protects the metal wall from corrosion. The exterior of the hydrator is cooled by water jackets, while the interior is protected by a flowing film of cooled, recycled fluid. The product is drawn off from the recycle-acid storage tank. The hydrator is also supplied with a spray of weak acid coming from the separator tower, which scrubs the off-gases from the hydrator free from phosphorus oxide gases. Phosphoric acid of a concentration 66%-83% P_2O_5 has been produced in this design.

Advantages

Process proven by TVA

Process owners

Tennessee Valley Authority

Engineering licensees

Davy Powergas Ltd
Gulf Design Company
Hitachi Zosen
The M. W. Kellogg Company
Occidental Chemical Company
Submerged Combustion (Engineering) Ltd
Wellman Incandescent Ltd
Woodall-Duckham Ltd

Cost of typical plant

A plant having a capacity of 200 t/d P_2O_5 from phosphate rock can cost $20 million to $22 million; a 300 t/d plant can cost $28 million to $30 million and a 2 x 300 t/d plant can cost $42 million to $46 million.

Single superphosphate (SSP)

Single superphosphate (SSP), also called normal, or ordinary, superphosphate, is the oldest source of phosphorus for fertilizer. It has a P_2O_5 equivalent of 16%-22%. The overall reaction of phosphate rock with sulphuric acid can be represented by the following equation:

$$Ca_{10}(PO_4)_6F_2 + 7H_2SO_4 \rightarrow 3Ca(H_2PO_4)_2 + 7CaSO_4 + 2HF$$

SSP may be manufactured in small, inexpensive plants, since the process is simple. The physical steps involved are (a) mixing the phosphate rock with the acid; (b) permitting the mix to attain a solid form, otherwise known as denning; and (c) storage to permit the acidulation reaction to go to completion, known as curing.

The phosphate rock is ground to 100-150 mesh and mixed with 50%-75% sulphuric acid in a specially designed vessel, either by batches or continuously. The standard batch mixer is of the pan type, with vertical rotating stirrers. As the pan turns in one direction, the blades rotate in the opposite direction. The usual mixing time is 2-3 minutes. Rock and acid are fed to the mixer simultaneously in a weighed ratio, about 0.38 t of 100% acid to 1 t of rock. The strength of the acid used is usually 70%, to yield a product with maximum content of monocalcium phosphate $(Ca(H_2PO_4)_2)$, the water-soluble component. Continuous mixing can be done in the TVA cone mixer, which is becoming more popular in modern plants.

The oldest type of den is known as the box den, from which the product is removed with a dragline. The more recent type of den is the Sturtevant den, where the product is removed by a cutter and conveying equipment. Continuous-process dens are also available, two of which are the Broadfield den, which has a slat-type conveyor and the Sackett den, which has a channel conveyor. A rotary den has also been developed.

After denning, the reacted mass is cured to enable the acidulation reaction to go to completion. In conventional processes curing is done in a shed for 3-4 weeks. The cured product is excavated, milled, screened and either bagged directly or steam granulated before bagging.

Advantages

  Simple process
  P_2O_5 in water-soluble form
  Low investment cost
  Free-flowing and non-caking granulated product

Disadvantages

  Low nutrient content of product

Process owners

  Fisons Ltd, Fertilizer Division
  Mitsubishi Chemical Industries Ltd
  Nordengren Patenter AB
  Tennessee Valley Authority
  Ugine Kuhlmann SA

Engineering licensees

  Davy Powergas Ltd
  Didier Engineering GmbH, Industrieanlagenbau
  Foster Wheeler Corporation
  Engineering Company for Inorganic Chemistry and Fertilizer Industry (IPRAN)
  Krebs et Cie SA
  Lurgi Chemie und Hüttentechnik GmbH
  The M. W. Kellogg Company
  Niigata Ryusan Company Ltd
  Norden Industrias Metalurgicas SA
  Sim-Chem Division, Simon-Carves Ltd
  Sumitomo Chemical Company Ltd
  Woodall-Duckham Ltd

Cost of typical plant

  A 400 t/d SSP plant will cost $1.5 million to $2.0 million. Plant cost includes rock-grinding facilities but excludes sulphuric acid facilities.

Single superphosphate (SSP)

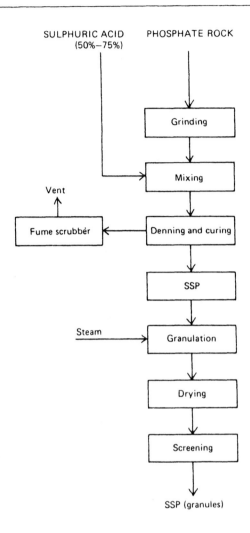

Manufacture of SSP

Triple superphosphate (TSP)

Triple superphosphate (TSP) is an excellent source of plant nutrient phosphorus; almost all of its P_2O_5 content of 44%-48% is in water-soluble form.

TSP is manufactured by reacting phosphoric acid (47%-54% P_2O_5) with phosphate rock:

$$Ca_{10}(PO_4)_6F_2 + 14H_3PO_4 \rightarrow 10Ca(H_2PO_4)_2 + 2HF$$

Phosphoric acid obtained from either the thermal or the wet-process can be used; however, the wet process is the usual choice. When the rock and phosphoric acid are mixed, the resulting mass solidifies much more rapidly than in the case of SSP. That is the main difference between the two processes.

The manufacture of TSP is described below.

Phosphate rock, ground to such size that 90% passes through 100 mesh, and 47%-54% P_2O_5 phosphoric acid are metered continuously into a mixer in a weight ratio of 1 to 1.5-1.6. The mixer employed is usually the TVA cone type, but one of the turbine-agitator type is also used. The slurry produced is discharged into a belt conveyor that acts as a den, where the mass hardens in a short time. The product is then broken by a disintegrator and sent to a curing pile. After a specified curing period (about four weeks), the product, which is in a pulverized form, is bagged.

The den system is provided with a fume-scrubbing arrangement, where a water wash is used to reduce atmospheric pollution by the fluorine-bearing off-gases.

TSP is manufactured in granular form also by using the conventional slurry process. Merchant-grade phosphoric acid (47%-54% P_2O_5) and phosphate rock are mixed in cascade reactors. The slurry from the last reactor is granulated either with a pug-mill type blunger or a drum-type granulator. The granules are screened to separate and recycle the oversize material and fines. The product is sent to storage.

TSP manufacture by a batch process is not popular since it presents difficulties in that a hard mass sets on acidulation and a surfactant is needed.

Advantages

    High P_2O_5 content, almost all in water-soluble form
    Pulverized product good for ammoniation
    Granulation plant flexible for ammonium phosphate production
    Good storage property of granulated product

Disadvantages

    Low level of sulphur (sometimes)
    Phosphoric acid supply or manufacturing facility needed, which means extra investment
    High cost of granulating plant

Process owners

    Fisons Ltd, Fertilizer Division
    Jacobs Engineering Company
    Nordengren Patenter AB
    Société de Prayon SA
    Tennessee Valley Authority
    Ugine Kuhlmann SA

Engineering licensees

    Davy Powergas Ltd
    Didier Engineering GmbH, Industrieanlagenbau
    Foster Wheeler Corporation
    Engineering Company for Inorganic Chemistry and Fertilizer Industry (IPRAN)
    Krebs et Cie SA
    Lurgi Chemie und Hüttentechnik GmbH
    Niigata Ryusan Company Ltd
    Sturtevant Mill Company
    VEB Chemie–Ingenieurbau Leipzig
    Woodall-Duckham Ltd

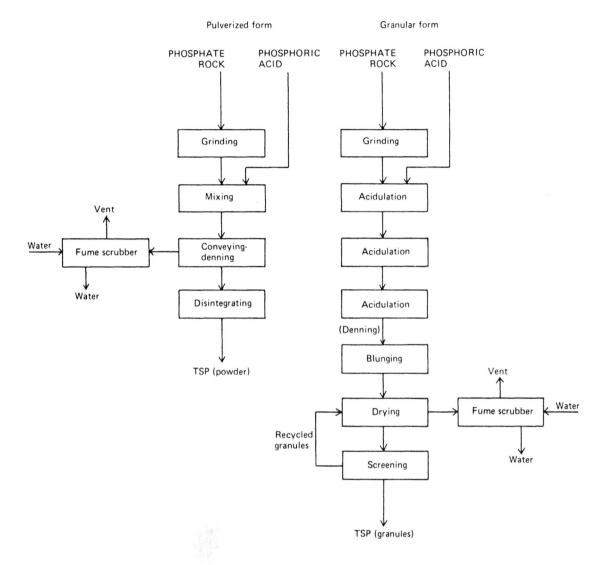

Manufacture of TSP

Cost of typical plant

A 500 t/d TSP plant, including the rock-grinding facilities but excluding a phosphoric acid plant, can cost $3.3 million for producing a pulverized product; a 1 000 t/d plant can cost $5.5 million to $6.0 million, a 1 900 t/d plant can cost $7.5 million to $8.0 million. A plant of an equal capacity for producing a granulated product can cost 2-3 times more.

Monoammonium phosphate (MAP)

Monoammonium phosphate (MAP) is the product of the reaction of phosphoric acid with ammonia:

$$NH_3 + H_3PO_4 \rightarrow NH_4H_2PO_4$$

It is itself a rich NP fertilizer, with 11%-12% N and 48%-53% P_2O_5, but is also used as an intermediate in the manufacture of high-analysis NPK fertilizers. MAP is produced in powdered or granular form. Two processes for the first and one for the second are described.

Fisons process (powder)

Anhydrous ammonia and wet-process phosphoric acid (45%-52% P_2O_5) are individually metered and injected in a mole ratio of 1 to 1 into a reactor kept under a pressure of 2 bar. The temperature inside the reactor is maintained at 170°C, so that the moisture content in the phosphoric acid is reduced to less than 10% after the reaction is over. The solution from the reactor, containing mostly MAP, is sprayed under its own pressure into the top of a natural-draft tower, where residual water is vaporized and MAP solidifies as the solution flows down the tower. The product, which is graded as 11-53-0, is removed continuously from the tower and sent to storage. The moisture in the product is 5%-7% and the grain size is 0.2-1.5 mm.

Advantages
  Proven process
  Ammonia conversion efficiency 96%-97%

Process owners
  Fisons Ltd, Fertilizer Division

Engineering licensees
  Davy Powergas Ltd
  Foster Wheeler Corporation
  Lurgi Chemie und Hüttentechnik GmbH

Cost of typical plant

A 400 t/d MAP plant can cost $1.4 million, excluding necessary off-site facilities. The cost can be taken as 75%-100% more if all facilities are included.

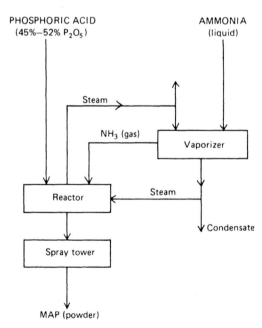

MAP by the Fisons process

PhoSAI process (powder)

Gaseous ammonia and dilute phosphoric acid are reacted in a stainless steel reactor to an N to P ratio of 1.3 to 1. Ammonia fumes escaping from the reactor are scrubbed by fresh phosphoric acid and returned to the reactor. The ammoniated slurry enters moisture-disengagement separators where more phosphoric acid is added to reduce the mole ratio to 1 to 1; owing to the heat of the reaction, further moisture is driven off. The product is screened to remove oversize particles and sent to storage. The oversize particles are crushed and returned. The product grade obtained is 12-50-0; moisture content, 6%; grain size, 0.2-1.5 mm.

Advantages

Process well known
Ammonia conversion efficiency 96%-97%

Process owners

Scottish Agricultural Industries Ltd

Engineering licensees

Sim-Chem Division, Simon-Carves Ltd
Woodall-Duckham Ltd

Cost of typical plant

A 400 t/d MAP plant can cost $1.2 million to $1.4 million, excluding off-site facilities. The cost can be taken as 75%-100% more if all facilities are included.

Slurry process (granules)

Phosphoric acid (50%-54% P_2O_5) is neutralized with ammonia in a series of tanks. The mole ratio of ammonia to acid is maintained at 1.3-1.4 to 1. Vapours escaping from the neutralizer are scrubbed with phosphoric acid and returned to the neutralizer. The ammoniated slurry goes to a blunger, where further acid is added to adjust the mole ratio to 1 to

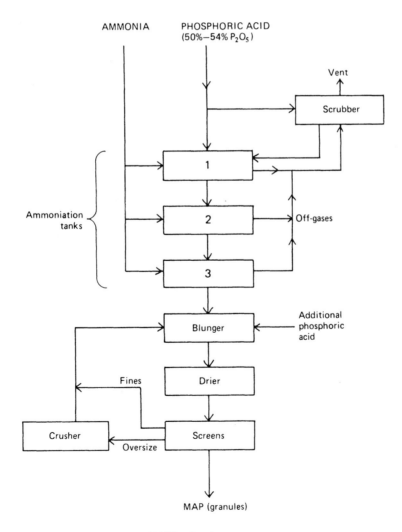

MAP by the slurry process

1. The granules from the blunger go to the drier and then to the screens. The oversize is crushed and returned to the blunger with the fines. The product from the drier is sent to storage.

Advantages

    Proven process

Disadvantages

    Recycle ratio high at 4-8 to 1
    High initial cost

Process owners

    Jacobs Engineering Company
    Mitsui Toatsu Chemicals, Inc.
    Stamicarbon BV
    Tennessee Valley Authority

Engineering licensees

    C. F. Braun and Company
    Chiyoda Chemical Engineering and Construction Company Ltd
    Coppée-Rust SA
    Didier Engineering GmbH
    Foster Wheeler Corporation
    Friedrich Uhde GmbH
    Jacobs Engineering Company
    Kellogg Continental BV
    Krebs et Cie SA
    Sim-Chem Division, Simon-Carves Ltd
    Woodall-Duckham Ltd

Cost of typical plant

A 400 t/d MAP plant can cost $20 million to $25 million for a battery limits plant. The cost will increase by 75%-100% to include off-site facilities and start-up cost.

Diammonium phosphate (DAP)

Diammonium phosphate (DAP) contains 2 moles of P_2O_5 per mole of ammonia. It is becoming more important in the family of NP fertilizers and is also used in formulations of high-analysis NPK fertilizers.

DAP formation can be represented as follows:

$$2NH_3 + H_3PO_4 \rightarrow (NH_4)_2HPO_4$$

Ammonia is reacted with merchant-grade phosphoric acid (50%-54% P_2O_5) in a pre-neutralizer operating at 115°C, where the mole ratio of ammonia to the acid is maintained at 1.4 to 1. (In the process owned by Dorr Oliver, Inc., a series of three reactors is used.) The slurry is pumped to the ammoniator-granulator with recycled fines from the back end of the plant. More ammonia is injected here to increase the mole ratio to 1.8-2.0 to 1. The product granules from the granulator go to a co-current drier to reduce the moisture content to 1% and then are screened. The oversize and fines are recycled; the ratio of recycle to fresh feed is 4-6 to 1. The product from the screens is cooled and sent to storage. Ammonia losses from the pre-neutralizer and the granulator are kept to the minimum by scrubbing the vapours with phosphoric acid and returning the acid to the pre-neutralizer. This process is known as the TVA process, and the grades obtained are 18-46-0 and 16-48-0.

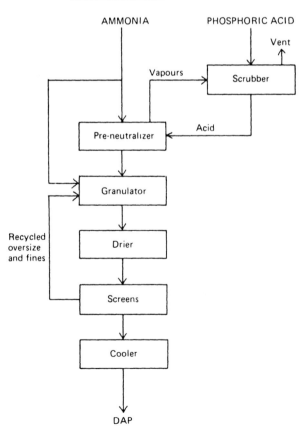

Manufacture of DAP

Advantages

    High-grade fertilizer obtained
    Good storage properties of product

Disadvantages

    For use as fertilizer intermediate, product has to be powdered before processing
    High initial and operating costs because of high recycle ratio

Process owners

    Dorr Oliver, Inc.
    Fisons Ltd, Fertilizer Division
    Stamicarbon BV
    Tennessee Valley Authority

Engineering licensees

    The Austin Company
    Continental Engineering (Ingenieursbureau voor de Procesindustrie NV)
    Coppée-Rust SA
    Davy Powergas Ltd
    Foster Wheeler Corporation
    Friedrich Uhde GmbH
    Humphreys and Glasgow Ltd
    Jacobs Engineering Company
    Kaltenbach et Cie SA
    Lurgi Chemie und Hüttentechnik GmbH
    Mitsubishi Chemical Industries Ltd
    Mitsui Toatsu Chemicals, Inc.
    Monsanto Enviro-Chem Systems, Inc.
    The M. W. Kellogg Company
    Sim-Chem Division, Simon-Carves Ltd
    Sumitomo Chemical Company Ltd
    Woodall-Duckham Ltd

Cost of typical plant

A 400 t/d DAP plant will cost $6.0 million. A 850 t/d plant will cost approximately $8.0 million. For a 1 000 t/d plant, the cost can go up to $12.0 million.

Ammonium sulphate phosphate

Ammonium sulphate phosphate fertilizer is composed of a mixture of ammonium sulphate $((NH_4)_2SO_4)$ and ammonium phosphate $(NH_4H_2PO_4)$.

A mixture of sulphuric and phosphoric acids is directly neutralized with ammonia. Phosphoric acid (30%-32% P_2O_5) is mixed with 98% sulphuric acid in a mole ratio P_2O_5/H_2SO_4 of 1 to 2. Ammonia is added at the rate of about 0.22 t per tonne of product. The slurry, containing 70%-75% solids, is granulated in a blunger. The product granules are dried and screened before they are transferred to storage. Any offsize particles are recycled. The product grade obtained is 16-20-0; provision can be made to add urea to the blunger so as to give a grade of 20-20-0. An ammonium sulphate solution can be used in place of sulphuric acid.

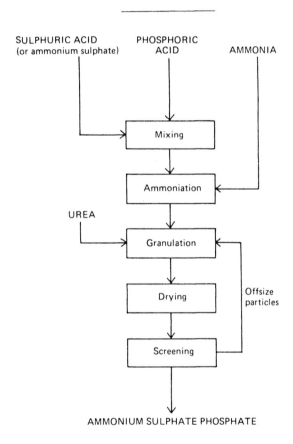

Manufacture of ammonium sulphate phosphate

Process owner

Dorr Oliver, Inc.
Fisons Ltd, Fertilizer Division
Stamicarbon BV

Engineering licensees

C. F. Braun and Company
Chiyoda Chemical Engineering and Construction Company Ltd

Coppée-Rust SA
Davy Powergas Ltd
Didier Engineering GmbH
Foster Wheeler Corporation
Friedrich Uhde GmbH
Humphreys and Glasgow Ltd
Jacobs Engineering Company
Kellogg Continental BV
Krebs et Cie SA

Lurgi Chemie und Hüttentechnik GmbH
San Kagaku Company Ltd
Sim-Chem Division, Simon-Carves Ltd
Stone and Webster Engineering Corporation

Cost of typical plant

A 1 000 t/d plant with necessary facilities can cost around $10 million.

Ammonium phosphate nitrate (with potash)

When ammonia is reacted with a mixture of phosphoric and nitric acids, a mixture of ammonium nitrate, MAP and DAP is formed.

Phosphoric acid (40% P_2O_5) and nitric acid of 55% strength are mixed in the required ratio, depending on the formulation grade required, and reacted with gaseous ammonia in the neutralizer. The resulting solution is sent to a vacuum evaporator, where the concentration is increased to about 99%. The temperature of the evaporator melt is 170°C, and the evaporator pressure is kept at about 0.5 bar. The nitrophosphate melt is then pumped to the top of the prilling tower and is sprayed. The solid prills are recovered from the bottom of the tower and cooled to 40°C in a rotating drum. The cooled product is screened and coated with kieselguhr. The fines and crushed oversize material are sent for re-mixing with the evaporator melt. Potash, in the form of potassium chloride (KCl), may also be added as required. The system is provided with facilities for recovering the ammoniacal vapours from the evaporator and the dust from the back end of the plant.

For grades having an N/P_2O_5 ratio below 0.8 and N/K_2O and P_2O_5/K_2O ratios below 0.6, a prilling process may not be suitable. In such cases, a pug-mill granulator is employed. The product from the pug mill is dried, cooled, screened and then coated before being sent to storage.

The grades of products manufactured in this process are typically 14-14-14 and 17-17-17.

Advantages

High-analysis fertilizer obtained
Low initial cost of plant

Disadvantages

Limited number of formulation grades obtainable
High product cost owing to high raw-material cost

Process owners

BASF Aktiengesellschaft
Fisons Ltd, Fertilizer Division
Nissan Chemical Industries Ltd
Stamicarbon BV

Engineering licensees

Chiyoda Chemical Engineering and Construction Company Ltd
Coppée-Rust SA
Davy Powergas Ltd
Foster Wheeler Corporation
Friedrich Uhde GmbH
Humphreys and Glasgow Ltd
Jacobs Engineering Company
Sim-Chem Division, Simon-Carves Ltd

Cost of typical plant

A 1 000 t/d plant with the necessary facilities can cost around $10 million.

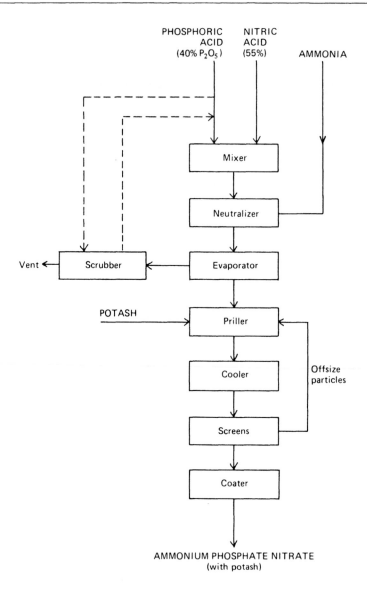

Manufacture of ammonium phosphate nitrate (with potash)

Urea ammonium phosphate (UAP)

Urea ammonium phosphate (UAP) fertilizers have high nutrient analyses and have therefore become very popular in the last decade all over the world. Urea (46% N), wet-process phosphoric acid (54% P_2O_5) and potash (KCl, 60% K_2O), along with ammonia, form the raw materials for making UAP fertilizer. Additionally, the phosphate present in the fertilizer is in the desirable water-soluble form.

The four manufacturing methods described here, although quite different in operation conditions, cost about the same – $8 million to $10 million for a 1 000 t/d plant.

Conventional slurry process

Almost all UAP plants in the world use the conventional slurry process. Wet-process phosphoric acid (48%-54% P_2O_5) is added to the pre-neutralizer, where anhydrous ammonia is injected to form ammonium phosphate slurry. Ammoniation is controlled to have a N_2/P_2O_5 ratio of 1.3-1.4. Under this condition, the slurry has 12% water and the ammonium phosphate solids have maximum solubility in it; at the same time, it is in a pumpable form in spite of the high concentration of solids. This control is very important; if too much water enters the granulator, the granulation and the final product quality will be adversely affected.

The MAP and DAP slurry is pumped and evenly distributed into a TVA-type drum granulator or a paddle blunger of the pug-mill type, over a bed of recycled material. This recycled material consists of prilled urea, undersize granules from the screen, crushed oversize from the mills and the fertilizer dust reclaimed from the scrubber. Also, KCl and filler are added to the recycle as per formulation requirements. Further ammoniation is carried out in the granulator or blunger to complete the ammoniation of the slurry to a mole ratio of 1.8-1.9. The granules coming out are dried in a co-current drier from a moisture content of 2.0%-2.5% to 1.0%. The dried product is sent to the screens, where the oversize and the fines are separated and recycled as necessary. The screened product goes to a cooler and a coating drum and then to storage.

The addition of ammonia as in the granulator has a beneficial effect on the granulation process. Apart from increasing the mole ratio, the heat of the reaction helps to remove some of the moisture and raises the granulator temperature.

A scrubber recovery system is provided to recover unreacted ammonia vapour from the pre-neutralizer and granulator and to filter the dust particles from the drier and cooler cyclones. Scrubbing is achieved by intimately contacting the ammonia and dust-laden air with a weak circulating phosphoric acid solution in a venturi-type scrubber, and the solution is fed back to the reactor.

In this process, UAP is commonly made in the following grades:

22-22-11
19-19-19
17-17-17
28-28-0
20-20-0
14-28-14
24-24-0

Granulation of UAP of the 28-28-0 and 20-20-0 formulations is generally favoured by higher temperature and lower moisture levels in the granulator. When potash is introduced into the mixture, the granulation conditions are more critical and have to be controlled closely, since the salts in the formulation are highly soluble, and the mass may become sticky instead of free-flowing.

Advantages

Maximum nutrient content
Practically all P_2O_5 in water-soluble form
Savings in bagging, transportation, storage, distribution and application

Disadvantages

High-potash grades need to be carefully controlled for quality
Low critical relative humidity for the product
Coating frequently required to maintain keeping qualities during storage

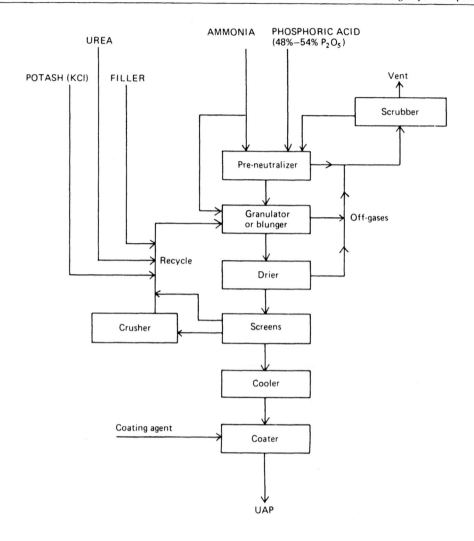

UAP by the conventional slurry process

High initial plant cost owing to large number of pieces of equipment
Dependent on raw-material supply from outside

Process owners

Dorr Oliver, Inc.
Fisons Ltd, Fertilizer Division
Mitsui Toatsu Chemicals, Inc.
Norsk Hydro AS
Stamicarbon BV
Tennessee Valley Authority

Engineering licensees

Coppée-Rust SA
Davy Powergas, Inc.
Didier Engineering GmbH
Fluor Corporation
Friedrich Uhde GmbH
H. K. Ferguson Company, Inc.
Humphreys and Glasgow Ltd
Lurgi Chemie und Hüttentechnik GmbH
The M. W. Kellogg Company
Sim-Chem Division, Simon-Carves Ltd

Cost of typical plant

For a plant of 1 000 t/d capacity of the 19-19-19 formulation, the cost will be approximately $8 million.

Melt process

The melt process used by TVA to make UAP is a new development in that it effectively uses the heat of the reaction between ammonia and phosphoric acid, and the drying step is avoided. A simple but highly effective pipe reactor is the key piece of equipment in this process.

Wet-process phosphoric acid (52%-54% P_2O_5) is metered to a spray absorber, where excess ammonia

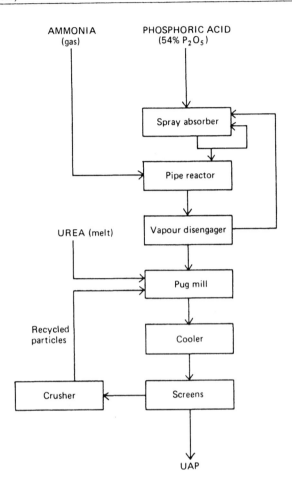

UAP by the melt process

from the pipe reactor (through the vapour-disengager equipment) is recovered. Phosphoric acid that has partially reacted in the absorber is recycled to the absorber, and a slip stream from recycle is taken to the pipe reactor, a pipe of type 316 stainless steel (schedule 40), 6 in. (15 cm) in diameter and about 10 ft (3 m) long. Phosphoric acid enters the pipe reactor through a side inlet, while anhydrous gaseous ammonia, pre-heated to 135°C, enters through the main inlet at a measured rate. The slurry goes to the vapour disengager, where free water and some combined water are evolved together with unreacted ammonia, which is recovered in the spray absorber as already mentioned. The slurry in the form of a melt is mixed by the helical blades of the disengager, and the compacted mass is fed to the double-shaft pug mill. Urea melt and recycle fines are also fed to the pug mill. The granules from the pug mill are cooled and screened and the product is sent to storage, while the oversize and undersize particles are recycled. Typical grades of UAP made in this process are 28-28-0 and 11-55-0.

Advantages

Drier and attendant capital and fuel cost eliminated

Polyphosphate content of 15%-20%, which improves storage quality

Disadvantages

Process still to be widely adopted

Process owners

Tennessee Valley Authority

Engineering licensees

Organizations outside the United States of America free to use process

Cost of typical plant

For a 1 000 t/d UAP plant, the cost will be $8 million to $10 million.

Dry-ingredients process

The dry-ingredients process, developed by Fisons to make UAP fertilizer, uses solid intermediates as feedstock, coupled with low-level ammoniation in the granulator.

The raw materials—urea, MAP and potash—along with the necessary filler, are fed to the plant through a preparation section where any oversize materials are crushed. The partially mixed raw materials and the recycle stream from the back end of the plant are fed to the granulator. The required quantity of low-pressure steam is fed into the rolling mass in the granulator by a sparger. Sulphuric acid and ammonia are also added to the granulator if required. The granulated material is discharged into a rotary co-current drier. The dried material is screened to remove oversize and fines. The product is cooled in a rotary cooler and then coated with finely divided kieselguhr, clay, or fuller's earth, using an oil adhesive.

The offsize materials from the screens are recycled to the granulator after the oversize has been crushed. The drier and cooler are provided with high-efficiency dust cyclones, and the separated dust is returned to the process.

In this process, UAP is commonly made in the following grades:

15-15-15
12-24-12
19-19-19
20-10-10

Advantages

Use of solid intermediates
Low recycle ratio, and hence higher throughput
Simpler scrubbing facilities owing to lower emission of fumes

Process owners

Fisons Ltd, Fertilizer Division

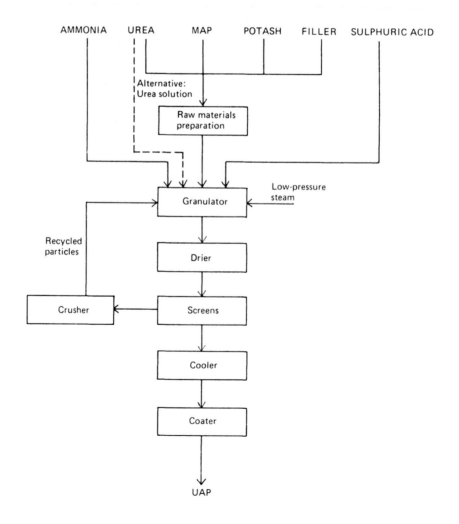

UAP by the dry-ingredients process

Engineering licensees

   Davy Powergas Ltd
   Foster Wheeler Corporation
   Lurgi Chemie und Hüttentechnik GmbH

Cost of typical plant

   A 1 000 t/d plant can cost about $10 million.

Melt oil-cooling process

The melt oil-cooling process developed by Mitsui Toatsu of Japan consists of mixing urea melt with ammonium phosphate and, if required, potassium chloride.

The ammonium phosphate used in the process is either solid MAP or phosphoric acid and ammonia reacted in a pre-neutralizer to form a slurry. Solid urea prills are taken through a melter before they flow into a mixing vessel, where ammonium phosphate and potassium chloride are added. The mixer also receives the recycled material from the back end of the plant.

The slurry from the mixture is kept at 102°-140°C so as to reduce the moisture content to 0.5%, and the mixing operation is subjected to close control to avoid decomposition of urea by condensation of ammonium phosphate into polyphosphate. The resulting suspension is further evaporated and prilled in specially designed equipment. The prills are directly admitted into a shallow tank containing the cooling oil. An oil depth of 2 m is maintained to solidify the prills completely up to 5 mm in diameter. The prills are then separated by sedimentation and centrifuging; after screening, the final product is coated with oil to 0.5%-1.0% by weight and sent to storage. The off-size material is returned to the mixer.

Advantages

   Expensive prilling tower and drying equipment eliminated; hence lower investment

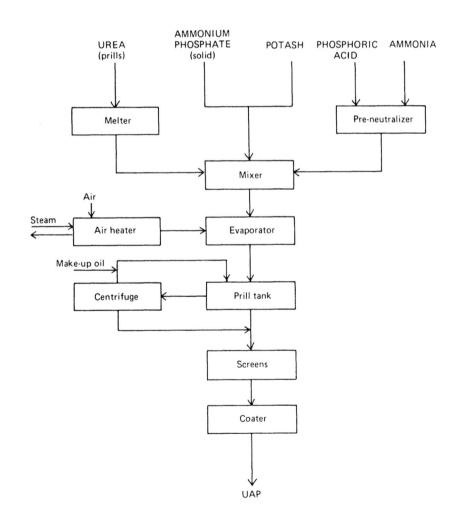

UAP by the melt oil-cooling process

Disadvantages

    Mixing conditions critical
    Process yet to be proven in commercial practice

Process owners

    Mitsui Toatsu Chemicals, Inc.

Engineering licensees

    Davy Powergas Ltd
    Fluor Corporation
    H. K. Ferguson Company, Inc.
    Lurgi Chemie und Hüttentechnik GmbH
    The M. W. Kellogg Company
    Toyo Engineering Corporation

Cost of typical plant

A 1 000 t/d capacity plant can cost about $8 million to $10 million.

Nitrophosphates

Nitrophosphates contain the fertilizer nutrients nitrogen and phosphorus. They are obtained by acidulation of phosphate rock with nitric acid; the resulting slurry contains a complex mixture of nitrates and phosphates.

The acidulation reaction between phosphate rock and nitric acid can be expressed as follows:

$$Ca_{10}(PO_4)_6F_2 + 14HNO_3 \rightarrow 3Ca(H_2PO_4)_2 + 7Ca(NO_3)_2 + 2HF$$

The phosphate component of the product is in the form of water-soluble monocalcium phosphate. However, the calcium nitrate makes monocalcium phosphate unstable, and the product is hygroscopic. Various processes have been developed to eliminate the calcium nitrate; the methods used can be summarized as follows:

(a) Removal by crystallization followed by centrifuging or filtration (the Odda process is an example of this method);

(b) Conversion to calcium sulphate by using ammonium sulphate or sulphuric acid (sulphonitric process);

(c) Conversion to monocalcium phosphate by using phosphoric acid (phosphonitric process);

(d) Conversion to calcium carbonate by using carbon dioxide (carbonitric process).

In all these processes, the nitrogen is restored by adding ammonia at some point. Potash may also be added to obtain a variety of NPK grades.

Odda process

Phosphate rock and 55%-60% nitric acid are fed to the first of the two digesters continuously. (An excess of nitric acid up to 10% is used). A small amount of defoamer is also added. The rock and acid mixture flows to the second digester, where dissolution is completed and the digested liquor containing monocalcium phosphate and calcium nitrate flows to a buffer tank. The slurry is then pumped to the crystallizers, which are equipped with cooling coils cooled by brine and operated in series, partly continuously and partly in batches. This operation provides an effect of continous processing of the slurry. The temperature maintained in the crystallizers is based on the amount of calcium nitrate to be removed and the degree of water solubility to be obtained in the phosphate product. For example, a temperature of $10°C$ in the crystallizers will remove 50% of the calcium nitrate, and the water solubility will be 30%-40%. Easily filterable crystals of uniform particle size are obtained; these are separated from the stream on rotary tandem filters. The liquor from the calcium nitrate filtration stage is pumped to the neutralization section, where gaseous ammonia is added. Neutralization is carried out in stages to avoid the viscous solutions that occur at intermediate pH values. Depending on the requirement of N and P_2O_5 in the final product, ammonium nitrate produced in the calcium nitrate conversion (see below) is added after due concentration.

The neutralized solution is then pumped to the evaporator section, where the water content is reduced to 0.5% to enable prilling. If the product is to be granulated, a water content of up to 4% can be present. Evaporation is done under vacuum in two stages in specially designed equipment. Scrubbing units are provided to remove fluorine and ammonia in the vapours released during evaporation. After the evaporation, the solution is either prilled or granulated. Potash is added as necessary, before prilling or during the granulation step. The dry, hot product is screened and oversize and undersize granules removed and recycled. The product is then cooled and coated to maintain quality during storage. The calcium nitrate by-product separated in the filter can either be used in the production of calcium ammonium nitrate fertilizer (CAN) or dehydrated and sold, or converted to ammonium nitrate and calcium carbonate.

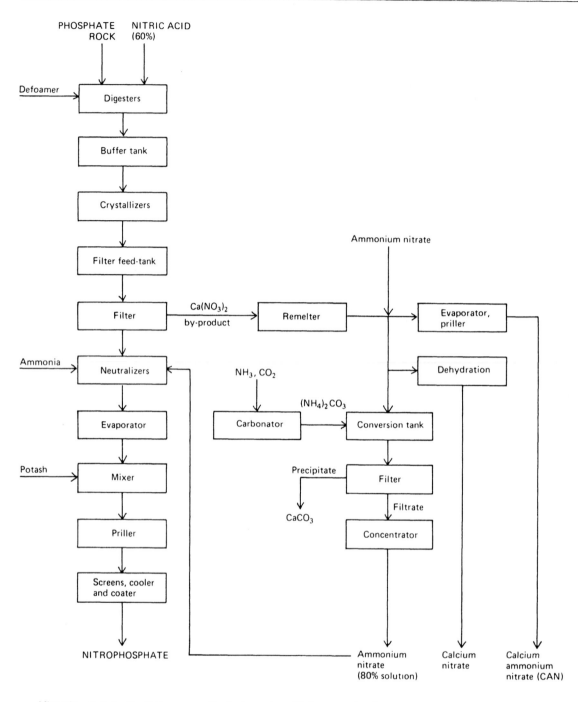

Nitrophosphate by the Odda process. Product may be prilled or granulated; only the prilling route is shown

CAN is obtained by treating calcium nitrate with ammonium nitrate and evaporating to 99%. The product is hygroscopic, but it can be prilled, flaked or conditioned by coating with oil to improve its storage properties.

Alternatively, the calcium nitrate can be converted to ammonium nitrate by treating it with ammonium carbonate. The slurry is filtered to remove the precipitated calcium carbonate, and the ammonium nitrate solution is concentrated to 80% and returned to the main stream (see above).

Nitrophosphate fertilizer is commonly made in the following grades by this process:

20-30-0
23-23-0
28-14-0
20-20-0
17-17-17
15-15-15
22-11-11

The water-soluble component of P_2O_5 is also related to the grade of the product. At 20-20-0 grade, the water solubility is about 30%, whereas at grades 23-23-0 and above, it can increase to 60%-70%.

Advantages

Only process to produce water-soluble nitrophosphate without using sulphuric or phosphoric acids
Fine grinding of rock not required
P_2O_5 recovery efficiency high at 99%
Reduced atmospheric pollution by fluoride

Disadvantages

N/P_2O_5 ratio of product too high for some uses
Cannot tolerate more than 5% silica in rock without pre-treatment
Cannot tolerate high concentrations of organic compounds in rock

Process owners

Chemische Fabrik Kalk GmbH
Norsk Hydro AS
Odda Smelteverk AS
Stamicarbon BV

Engineering licensees

Chiyoda Chemical Engineering and Construction Company Ltd
Coppée-Rust SA
Davy Powergas Ltd
Didier Engineering GmbH
Foster Wheeler Corporation
Friedrich Uhde GmbH
Humphreys and Glasgow Ltd
Kellogg Continental BV
Lurgi Chemie und Hüttentechnik GmbH
Sim-Chem Division, Simon-Carves Ltd
Stone and Webster Engineering Corporation
Toyo Engineering Corporation

Cost of typical plant

A plant having capacity of 1 000 t/d of 23-23-0 nitrophosphate and 950 t/d of CAN costs about $26 million to $30 million, including rock storage and auxiliary facilities, but excluding nitric acid plant.

Sulphonitric process

In the sulphonitric process, coarse-ground rock passing through 10 mesh is acidulated with 53%-57% nitric acid, the temperature of the reactor being controlled around 65°C. The reaction is carried out in several tanks. The acidulated slurry is reacted further with ammonia and ammonium sulphate solution. The following reaction takes place:

$$Ca(NO_3)_2 \cdot 4H_2O + (NH_4)_2SO_4 \rightarrow 2NH_4NO_3 + CaSO_4 \cdot 2H_2O + 2H_2O$$

The resulting slurry containing ammonium nitrate and calcium sulphate, along with products of nitric acid acidulation, is sent for granulation, drying, screening and cooling. The oversize fraction is crushed and returned as recycle to be granulated with the fines. The cooled product is coated before being sent to storage.

In one variant of the process, part of ammoniation is done in the neutralizer and the remaining ammonia applied in the rotary drum granulator. In another variant, gypsum precipitated out as crystals is separated by means of a tilting-pan filter, similar to the one used in phosphoric acid plants.

The gypsum cake, after being washed to recover the mother liquor, is rewashed on a rotary vacuum filter. The resulting filter cake is fed to the ammonium sulphate recovery section. Gypsum is reacted here with ammonium carbonate by the Mersiburg process. The ammonium sulphate solution is recycled to the precipitation section.

The slurry from the filtration section, which contains ammonium nitrate and the phosphate mixture, is neutralized with ammonia and sent for evaporation. The concentration is done in two stages of evaporation and under vacuum, so as to get a moisture content of 1.0% in the product. The concentrated melt is sent for prilling; potassium chloride is added as required. The prills are cooled, screened and coated. Oversize and fine particles are recycled.

It is possible to obtain a product that has a water solubility of up to 70%. However, the best values obtained in practice are between 30% and 50%.

The product is usually made in the following grades in this process:

14-14-0
11-11-11
20-10-0
15-15-0

Advantages

Reasonable amount of water solubility in product
Sulphate in the product can be micronutrient

Process owners

PEC-Engineering
Pierrefitte Auby SA
Produits Chimiques Pechiny-Saint-Gobain
Stamicarbon BV
Tennessee Valley Authority

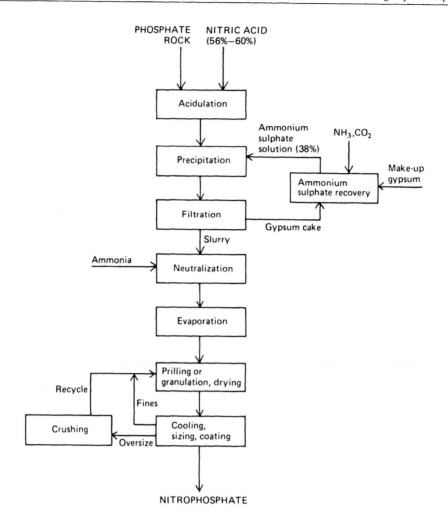

Nitrophosphate by the sulphonitric process

Engineering licensees

Continental Engineering
Coppée-Rust SA
Didier Engineering GmbH
Foster Wheeler Corporation
Friedrich Uhde GmbH
The M. W. Kellogg Company
PEC-Engineering
Sim-Chem Division, Simon-Carves Ltd
VEB Chemie-Ingenieurbau Leipzig

Cost of typical plant

A plant making 1 000 t/d of 15-15-0 fertilizer can cost $22 million to $26 million.

Phosphonitric process

Phosphate rock is acidulated with 50%-60% nitric acid, the temperature of the reactants being maintained at 70°-80°C. The calcium in the phosphate rock is converted into calcium nitrate, and after acidulation, water-soluble P_2O_5 from other sources, such as phosphoric acid, DAP or MAP, is added. Ammonia vapour at a temperature of 60°-70°C is then injected into the slurry for ammoniation. During ammoniation, the calcium nitrate produced by nitric acid digestion reacts with phosphoric acid to produce dicalcium phosphate, and ammonium nitrate is also formed. Thus this process step overcomes the calcium nitrate problem. More phosphoric acid is added than necessary to form dicalcium phosphate, so that water-soluble monocalcium phosphate is formed. The resulting slurry, if required, is combined with a potassium salt and is sent for granulation and drying. A spray-drying step, with a spherodizer, may be used instead. (In one variant of the process, two reactors and two ammoniation vessels are used for simpler process control, followed by spherodizer.) The product is screened, cooled and coated before being sent to storage. The oversize is crushed and recycled for granulation along with the fines.

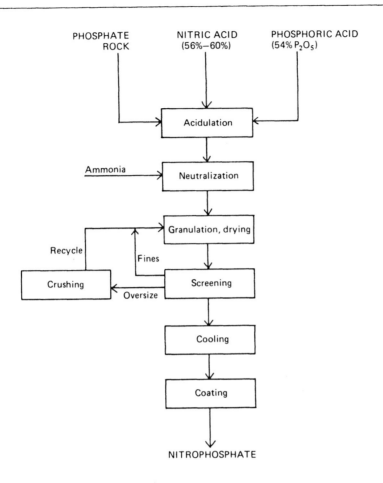

Nitrophosphate by the phosphonitric process

The product is commonly made in the following grades in this process; about 60% of the phosphates are present in water-soluble form:

16-23-0
14-14-14
17-35-0
20-20-0

Advantages

High N/P_2O_5 grades manufactured with reasonable degree of water solubility

Disadvantages

Dependent on availability of phosphoric acid

Process owners

PEC-Engineering
Produits Chimiques Pechiney-Saint-Gobain
Tennessee Valley Authority

Engineering licensees

Continental Engineering
Didier Engineering GmbH
Foster Wheeler Corporation
Friedrich Uhde GmbH
The M. W. Kellogg Company
PEC-Engineering
Sim-Chem Division, Simon-Carves Ltd
VEB Chemie-Ingenieurbau Leipzig

Cost of typical plant

A plant of 1 000 t/d capacity of 14-14-14 fertilizer can cost $22 million to $26 million, excluding the cost of a phosphoric acid plant.

Carbonitric process

In the carbonitric process, carbon dioxide (CO_2) is used to convert calcium nitrate into calcium carbonate. Acidulation proceeds as in the other nitrophosphate processes. After that, a stabilizer like magnesium or potassium sulphate is added to avoid reversion of dicalcium phosphate into water-insoluble tricalcium phosphate; the amount of the stabilizer required per tonne of product is 20-30 kg. After acidulation and stabilization, ammonia is added to neutralize the acid, and then the slurry is carbonated. The final pH should be maintained between 7.0 and 8.0 by adjusting the ammonia feed. In the product, phosphates are present in insoluble form up to 6%. The reaction slurry is granulated with recycling fines in a paddle mixer, dried and cooled. Potash may be added before granulation as necessary. The typical grades of the product are 16-14-0 and 13-11-12.

Advantages

Use of CO_2 to tie up calcium in calcium nitrate, as CO_2 is usually available as excess by-product in ammonia plants

Disadvantages

The phosphate is in water-insoluble form; 95% in citrate-soluble form. Hence, product application limited to certain crops and soils

Calcium carbonate in product will neutralize soil acidity, which is necessary to make dicalcium phosphate available

Process owners
PEC-Engineering

Engineering licensees
PEC-Engineering

Cost of typical plant

A plant of 1 000 t/d capacity can cost about $20 million.

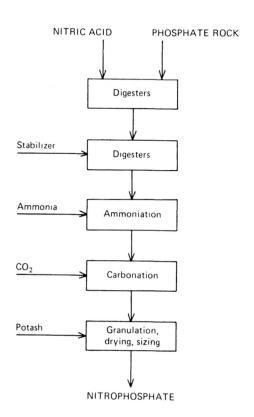

Nitrophosphate by the carbonitric process

Annex

LIST OF FIRMS

AB Celleco
　P.O. Box 94
　S-14700 TUMBA
　Sweden

Albright and Wilson Ltd
　1 Knightsbridge Green
　LONDON S.W.1
　United Kingdom

A. P. V. Kestner Ltd
　GREENHITHE
　Kent
　United Kingdom

The Austin Company
　3650 Mayfield Road
　CLEVELAND, Ohio 44121
　United States of America

Azote et Produits Chimiques (APC) SA
　143, route d'Espagne
　F-31000 TOULOUSE
　France

BASF Aktiengesellschaft
　Carl Bosch-Strasse 38
　D-6700 LUDWIGSHAFEN/RHEIN
　Federal Republic of Germany

Büttner Scheide Haas GmbH
　KREFELD-UERDINGEN
　Federal Republic of Germany

Canadian Industries Ltd
　P.O. Box 5201
　201 Queens Avenue
　LONDON, Ontario
　Canada N6A 4L6

C. F. Braun and Company
　ALHAMBRA, California 91802
　United States of America

Chemical Construction Corporation (Chemico)
　1 Penn Plaza
　NEW YORK, New York 10001
　United States of America

Chemie Linz AG
　Postfach 296
　A-4021 LINZ
　Austria

Chemische Fabrik Kalk GmbH
　Postfach 910210
　D-5000 COLOGNE 91
　Federal Republic of Germany

Chiyoda Chemical Engineering and Construction Company Ltd
　1850 Tsurumi-cho
　YOKOHAMA 230
　Japan

COCEI—Compagnie Centrale d'Etudes Industrielles
　44, avenue de Chatou
　F-92504 RUEIL-MALMAISON
　France

Collier Carbon and Chemical Corporation
　Union Oil Center
　P.O. Box 60455
　West Olympic Boulevard
　LOS ANGELES, California
　United States of America

Continental Engineering
　(Ingenieursbureau voor de Procesindustrie NV)
　Postbus 5294
　De Boelelaan 589
　AMSTERDAM
　Netherlands

Coppée-Rust SA
　251, avenue Louise
　B-1050 BRUSSELS
　Belgium

Davy Powergas GmbH
　Postfach 450280
　Aachenerstrasse 958
　D-5000 COLOGNE 41
　Federal Republic of Germany

Davy Powergas, Inc.
　P.O. Box 2436
　LAKELAND, Florida 33803
　United States of America

Davy Powergas Ltd
　8 Baker Street
　LONDON W.1M 10A
　United Kingdom

Didier Engineering GmbH
　Industrieanlagenbau
　P.O. Box 360
　Alpenstrasse 28
　D-4300 ESSEN
　Federal Republic of Germany

Dorr Oliver, Inc.
　77 Havemeyer Lane
　STANFORD, Connecticut 06094
　United States of America

Engineering Company for Inorganic Chemistry and Fertilizer Industry (IPRAN)
19-21 Mihail Eminescu Street
BUCHAREST
Romania

Farbenfabriken Bayer AG
D-5090 LEVERKUSEN-BAYERWERK
Federal Republic of Germany

The Fertilizers and Chemicals Travencore Ltd
UDYOGAMANDAL, Cochin
India

Fisons Ltd
Fertilizer Division
Harvest House
FELIXSTOWE, Suffolk IP11 7LP
United Kingdom

Fluor Corporation
2500 South Atlantic Boulevard
LOS ANGELES, California 90040
United States of America

Foster Wheeler Corporation
110 South Orange Avenue
LIVINGSTON, New Jersey 07039
United States of America

Fried. Krupp GmbH
Postfach 979
D-4300 ESSEN 1
Federal Republic of Germany

Friedrich Uhde GmbH
Deggingstrasse 10-12
D-4600 DORTMUND
Federal Republic of Germany

Gulf Design Company
Division of the Badger Company, Inc.
1052 East Memorial Boulevard
LAKELAND, Florida 33801
United States of America

Haines and Associates, Inc.
5050 Westheimer
HOUSTON, Texas 77027
United States of America

Heurtey Industries
B.P. 323
30, rue Guersant
F-75823 PARIS cedex 17
France

Hitachi Zosen
1-1, 1-chome Hototsubashi
Chiyoda-ku
TOKYO 100
Japan

H. K. Ferguson Company, Inc.
1 Erieview Plaza
CLEVELAND, Ohio 44114
United States of America

Humphreys & Glasgow Ltd
22 Carlisle Place
LONDON, SW1P 1JA
United Kingdom

IMI-Institute for Research and Development
P.O. Box 313
HAIFA
Israel

Jacobs Engineering Company
837 South Fair Oaks Avenue
PASADENA, California 91105
United States of America

Kaltenbach et Cie SA
192, Grandu Rue
F-92310 SEVRES
France

Kellogg Continental BV
P.O. Box 5294
De Boelelaan 873
AMSTERDAM
Netherlands

Kemira Oy
P.O. Box 330
Malminkatu 30
SF-00100 HELSINKI 10
Finland

Krebs et Cie SA
61, rue Pouchet
F-75017 PARIS
France

The Lummus Company
1515 Broad Street
BLOOMFIELD, New Jersey 07003
United States of America

Lurgi Chemie und Hüttentechnik GmbH
Gervinusstrasse 17/19
D-6000 FRANKFURT/MAIN
Federal Republic of Germany

Matthew Hall Engineering Ltd
Matthew Hall House
101-108 Tottenham Court Road
LONDON W.1A 1BT
United Kingdom

Mitsubishi Chemical Industries Ltd
c/o Mitsubishi Building
5-2, 2-chome Marunaichi
Chiyoda-ku
TOKYO 100
Japan

Mitsui Toatsu Chemicals, Inc.
3-2-5 Kasumigaseki
Chiyoda-ku
TOKYO
Japan

Monsanto Enviro-Chem Systems, Inc.
10 South Riverside Plaza
CHICAGO, Illinois 60606
United States of America

Montedison SpA
Largo Donegani 1-2
I-20134 MILAN
Italy

Annex — List of firms

The M. W. Kellogg Company
   1300 Three Greenway Plaza East
   HOUSTON, Texas 77046
   United States of America

Niigata Ryusan Company Ltd
   1686 Sebiya
   NIIGATA
   Japan

Nissan Chemical Industries Ltd
   Kowa-Hitotsubashi Building
   7-1, 3-chome
   Kanda-Nishiki-cho
   Chiyoda-ku
   TOKYO
   Japan

Nordengren Patenter AB
   LANDSKRONA
   Sweden

Norden Industries Metalurgicas SA
   Caixa Postal 391
   Av. Brig. Luiz Antonio 849
   SÃO PAULO
   Brazil

Norsk Hydro AS
   Bygdøy allé 2
   OSLO 2
   Norway

Occidental Chemical Company
   P.O. Box 1185
   2000 South Post Oak Road
   HOUSTON, Texas 77001
   United States of America

Odda Smelteverk AS
   N-5751 ODDA
   Norway

Outo Kumpu Oy
   P.O. Box 10280
   HELSINKI 10
   Finland

PEC-Engineering
   62/68 Rue Jeanne d'Arc
   F-75646 PARIS cedex 13
   France

Pierrefitte-Auby SA
   4, avenue Velasquez
   F-75008 PARIS
   France

Produits Chimiques Pechiney-Saint-Gobain
   B.P. 122
   63, rue de Villiers
   NEUILLY-SUR-SEINE
   France

Rhone-Poulenc Group
   Société des Usines Chimiques
   B.P. 753
   22, avenue Montaigne
   F-75008 PARIS
   France

San Kagaku Company Ltd
   1-9, 8-chome Hatscho-Bori
   Chuo-ku
   TOKYO
   Japan

Scottish Agricultural Industries Ltd
   25 Ravelston Terrace
   EDINBURGH EH4 3ET
   Scotland

Sim-Chem Division
   Simon-Carves Ltd
   STOCKPORT, Cheshire SK3 0RZ
   United Kingdom

Singmaster and Breyer
   100 Park Avenue
   NEW YORK, New York 10017
   United States of America

Société de Prayon SA
   ENGIS
   Belgium

Société Produits Chimiques et Synthesis SA
   29, rue Emile Zola
   BEZONS, Seine et Oisse
   France

Stamicarbon BV
   P.O. Box 10
   GELEEN
   Netherlands

Stone Webster Engineering Corporation
   90 Broad Street
   NEW YORK, New York 10004
   United States of America

Sturtevant Mill Company
   103 Clayton Street
   BOSTON, Massachusetts 02122
   United States of America

Submerged Combustion (Engineering) Ltd
   Harcourt
   HALESFIELD, Telford, Shropshire TF7 4NG
   United Kingdom

Sumitomo Chemical Company Ltd
   15, 5-chome Kitahama
   Higashi-ku
   OSAKA
   Japan

Swenson Evaporator Company
   Whiting Corporation
   HARVEY, Illinois 60426
   United States of America

Swindler Dressler Company
   441 Smithfield Street
   PITTSBURGH, Pennsylvania 15222
   United States of America

Tennessee Valley Authority (TVA)
   MUSCLE SHOALS, Alabama 35660
   United States of America

Toyo Engineering Corporation
   2-5, 3-chome Kasumigaseki
   Chiyoda-ku
   TOKYO
   Japan

Ugine Kuhlmann SA
  25, boulevard de l'Amiral Bruix
  F-75782 PARIS cedex 16
  France

VEB Chemieanlagenbau Magdeburg
  Schwiesausstrasse 6
  DDR-3018 MAGDEBURG
  German Democratic Republic

VEB Chemie-Ingenieurbau Leipzig
  Georgiring 1/3
  DDR-701 LEIPZIG
  German Democratic Republic

Wellman Incandescent Ltd
  Chemical Engineering Division
  Cornwall Road
  WARLEY, Worcestershire
  United Kingdom

Whiting Corporation
  Swenson Division
  15613 Lathrop Avenue
  HARVEY, Illinois 60426
  United States of America

Woodall-Duckham Ltd
  CRAWLEY, Sussex RH10 1UX
  United Kingdom

Bibliography

General

British Sulphur Corporation. Techno-economic appraisal for fertilizer industry in Arab states. London, 1976.

———— World guide to fertilizer processes and constructors. 5. ed. London, 1974.

Fertiliser Association of India. Fertiliser handbook on technology. New Delhi, 1975.

Sauchelli, V., ed. Chemistry and technology of fertilizers. New York, Reinhold, 1964.

———— Technology of phosphatic fertilizer.

Tennessee Valley Authority. World fertilizer market review and outlook. Muscle Shoals, Alabama, 1974.
  Report to the United States Agency for International Development.

United Nations Industrial Development Organization. Fertilizer manual. 1967. (ST/CID/15)
  Sales no : 67.II.B.1.

Sulphuric acid

Chari, K. S. Recent advances in sulphuric acid production. *Chemical age of India* (Bombay) 1976.

Hignett, T. P. Technical and economic comparison of sulphuric and nitric acid routes to phosphate fertilizers. Paper prepared for the ANDA/ISMA Seminar, Brazil, April 1975.

Sulphuric acid synthesis at elevated pressure. *Sulphur* (London) 124, May 1976.

United Nations Industrial Development Organization. The reduction of sulphur needs in fertilizer manufacture. (ID/SER.F/3)
  Sales no.: 69.II.B.26.

Phosphoric and superphosphoric acids

Hignett, T. P. Technical and economic comparison of nitric and sulphuric acid routes to phosphate fertilizers. Paper prepared for the ANDA/ISMA Seminar, Brazil, April 1975.

Nayar, K. P. A. *and* S. Balasubramanian. Wet process acid manufacture. Paper prepared for the FAI/ISMA Seminar, New Delhi, 1975

Roy, B. B. *and* A. K. Roy. Process for direct production of high concentration phosphoric acid. *F.A.I. technical review* (New Delhi) 15, 1973.

Slack, A. V., ed. Phosphoric acid. 2 vols. New York. Marcel Deccer, 1968 (Fertilizer science and technology series No. 1)

United Nations Industrial Development Organization. New process for the production of phosphatic fertilizers using hydrochloric acid. (ID/SER.F/5)
  Sales no.: 69.II.B.23.

———— The reduction of sulphur needs in fertilizer manufacture. (ID/SER.F/3)
  Sales no : 69.II.B.26.

Single superphosphate

Gupta, R. K. NPK fertilizers using urea and superphosphate. Paper prepared for the FAI/ISMA Seminar, New Delhi, 1975.

Ammonium sulphate phosphate

Sim-Chem. Data sheet 211-1,4. Stockport, Cheshire, United Kingdom.

Ammonium phosphate nitrate

Hunter *and* Hawksley. High analysis NPK fertilizer. Paper prepared for the FAI/ISMA Seminar, New Delhi, 1975.

Urea ammonium phosphate

Gupta, V. R. R. *and* P. N. Arunachalam. Production of complex fertilizers in Madras. Paper prepared for the FAI/ISMA Seminar, New Delhi, 1975.

Hock, C. Granulation of urea compound fertilizers. Paper prepared for the FAI/ISMA Seminar, New Delhi, 1975.

Manufacture of granular compound fertilizers based on urea. *By* J. D. C. Hemsley *and others*. Paper prepared for the FAI/ISMA Seminar, New Delhi, 1975.

Nitrophosphate

Agarwall, M. N. NP and NPK fertilizers through nitrophosphate route. Paper prepared for the FAI/ISMA Seminar, New Delhi, 1975.

Grundt, T. *and* I. S. Mangat. Modern nitrophosphate technology. Paper prepared for the FAI/ISMA Seminar, New Delhi, 1975.

Hignett, T. P. Technical and economic comparison of nitric and sulphuric acid routes to phosphate fertilizers. Paper prepared for the ANDA/ISMA Seminar, Brazil, April 1975.

Keleti, C. *and* E. Pelitti. Particular application of spherodizer in the Guano-Werke process. *Nitrogen* (London) March/April 1970.

United Nations Industrial Development Organization. Recent developments in the fertilizer industry. Report of the Second Interregional Fertilizer Symposium. (ID/94)
Sales no.: 72.II.B.31.

Van Luyt, P. Technical and economic comparison of precipitation and crystallization routes. Paper prepared for the FAI/ISMA Seminar, New Delhi, 1975.